FOREWOR

643.6 M39H 1980

REPAIR-MASTER FOR HAMILTON
AUTOMATIC DRYERS

This Repair Master contains information and service procedures to assist the service technician in correcting conditions that are not always obvious.

A thorough knowledge of the functional operation of the many component parts used on dishwashers is important to the serviceman, if he is to make a proper diagnosis when a malfunction of any part occurs.

We have used many representative illustrations, diagrams and photographs to portray more clearly these various components for a better over-all understanding of their use and operation.

IMPORTANT SAFETY NOTICE

You should be aware that all major appliances are complex electromechanical devices. Master Publication's REPAIR MASTER® Service Publications are intended for use by individuals possessing adequate backgrounds of electronic, electrical and mechanical experience. Any attempt to repair a major appliance may result in personal injury and property damage. Master Publications cannot be responsible for the interpretation of its service publications, nor can it assume any libility in connection with their use.

SAFE SERVICING PRACTICES

To preclude the possibility of resultant personal injury in the form of electrical shock, cuts, abrasions or burns, etc., that can occur spontaneously to the individual while attempting to repair or service the appliance; or may occur at a later time to any individual in the household who may come in contact with the appliance, Safe Servicing Practices must be observed. Also property damage, resulting from fire, flood, etc., can occur immediately or at a later time as a result of attempting to repair or service — unless safe service practices are observed.

The following are examples, but without limitation, of such safe practices:

1. Before servicing, always disconnect the source of electrical power to the appliance by removing the product's electrical plug from the wall receptacle, or by removing the fuse or tripping the circuit breaker to OFF in the branch circuit servicing the product.

NOTE: If a specific diagnostic check requires electrical power to be applied such as for a voltage or amperage measurements, reconnect electrical power only for time required for specific check, and disconnect power immediately thereafter. During any such check, ensure no other conductive parts, panels or yourself come into contact with any exposed current carrying metal parts.

2. Never bypass or interfere with the proper operation of any feature, part, or device engineered into the appliance.

3. If a replacement part is required, use the specified manufacturers part, or an equivalent which will provide comparable performance.

4. Before reconnecting the electrical power service to the appliance — be sure that:

    a. All electrical connections within the appliance are correctly and securely connected.
    b. All electrical harness leads are properly dressed and secured away from sharp edges, high-temperature components such as resistors, heaters, etc., and moving parts.
    c. Any uninsulated current-carrying metal parts are secured and spaced adequately from all non-current carrying metal parts.
    d. All electrical ground, both external and internal to the product are correctly and securely connected.
    e. All water connections are properly tightened.
    f. All panels and covers are properly and securely reassembled.

5. Do not attempt an appliance repair if you have any doubts as to your ability to complete it in a safe and satisfactory manner.

MASTER PUBLICATIONS

TABLE OF CONTENTS

SECTION 1

SERVICE CHECK LIST

The following diagnosis chart is intended to be only a starting point in proceeding with the servicing of automatic dryers. The diagnosis chart can only deal in generalities; to effectively service any appliance, the serviceman must thoroughly understand the mechanical functions and electrical circuitry of the appliance.

A considerable amount of time and money can be saved if a serviceman will take time to analyze the probable cause of a malfunction of a machine before proceeding to remove any parts. Always be sure first that the machine is properly installed and its power cord is plugged into a live receptacle that is properly fused. When checking electric dryers connected to 220 volts, be sure BOTH fuses are good. Be sure the gas is turned on and air is bled from the line when checking gas dryers. Check for proper air flow and make sure the operator has properly set the controls.

Always make a visual check first before using any testing equipment such as test lamps, voltmeters or ohmmeters. Before attempting to remove any electrical part from the machine, disconnect the power cord from the live receptacle. If a voltmeter or test lamp is being used for testing, the power cord must be plugged into a live receptacle, however.

SERVICE CHART (ALL DRYERS)

SERVICE PROBLEMS	DIAGNOSIS	REMEDY
The motor runs but the drum will not operate.	Broken or loose belt.	Replace or tighten belt.
	Loose motor or idler pulley.	Position and tighten pulleys.
	Frozen drum shaft.	Clean the shaft, replace bearings if necessary lubricate.
Drum operates but is noisy..	Drum warped.	Replace drum.
	Lower drum glide sticking or out of place.	Reposition or replace mechanical lower drum glide.
	Idler pulley noisy.	Lubricate idler arm pulley.
	Belt squeaking.	Use bar soap to lubricate the outer surface of the belt, or replace belt.
	Foreign objects are in the drum.	Remove all foreign objects that might be in the drum.
	Belt is frayed.	Replace the belt.
	Drum support bearings are worn, or need lubrication.	Lubricate the bearings and felt wick or replace the bearing. Mechanical, drum shaft and bearings, see text.
	Motor pulley loose.	Position and tighten the pulley set screw.
	The front or rear drum seals or bearings are worn out.	Replace seals or drum bearing.
	Machine not level or all leveling legs are not hitting the floor.	Level and adjust leveling legs.
Motor will not stop.	Incorrect wiring.	Check wiring diagram.
	Grounded motor or wiring.	Check motor, check other components for shorts to ground.
	Faulty timer.	Check timer, see electrical, timer text.
Motor does not start.	The fuse is out.	Replace blown fuse. Use 30 amp fuse.
	Timer is inoperative.	Check timer, see electrical, timers text.
	Motor is inoperative.	Check motor, see electrical motors text.
	Dryer is not properly connected.	Be sure the dryer is connected correctly, check voltage.
	Poor door switch connection, or an inoperative door switch.	Test the door switch circuit connections as explained in electrical, door switch text.
	Loading door is not closed properly.	Check the door to make sure the strike is in proper position.
Interior light does not burn.	Incandescent or ozone is burned out.	Check bulbs and replace if necessary. CAUTION: DO NOT CHECK OZONE BULB WITH 115 VOLT SOURCE.
	Door switch connections are poor or inoperative.	See electrical, door switch text.
	Inoperative light socket.	Replace the socket.

SERVICE CHART (ALL DRYERS) Continued

SERVICE PROBLEMS	DIAGNOSIS	REMEDY
Dryer smokes.	Lint has accumulated in the dryer.	Remove all lint.
	Wire insulation is burning.	Check and correct a short circuit. Replace or properly position the wiring.
	Overheated motor.	Check motor for lint build up and air circulation around motor.
Clothes drying too slowly.	A blocked lint trap or vent pipe.	Clean the lint trap and vent system.
	Vent pipe too long.	See installation instructions.
	Temperature control thermostat is set too low.	Check the blower temperature at which the thermostat clicks off. This should be around 165°F at the outlet of the lint trap. The thermostat takes 4 to 5 minutes to re-set with the blower running.
	Heat selector switch defective.	Check program switch.
	Safety thermostat is tripping.	This thermostat should not trip off in normal operation. This should only operate when the drum or blower stops. With the exhaust blocked, the dryer heat should cycle off after not more than 3 minutes of operation. Replace the thermostat if it trips in normal operation.
	House voltage is fluctuating or low.	Call the power company.
Clothes are not drying on automatic setting (for dryers with auto-dry only).	Timer advances to off too soon.	Set timer for more time on auto-dry position.
	Open control thermostat to heat source.	Check thermostat, see electrical control thermostat text.
Dryer does not dry properly.	Clothes are too wet when placed in the dryer — the dryer is overloaded.	Check washer for proper water extraction. Instruct the customer on load size, dry weight, detergent consumption.
	The blower assembly is plugged.	Remove the lint.
	Excessive lint accumulation in lint trap.	Remove accumulated lint from trap.
	Lint accumulation on blower guard.	Hinge blower guard open and clean.
	The cylinder rotates in reverse.	The cylinder (drum) rotates clockwise (facing front of dryer.) Motor windings are hooked up backwards if the dryer drum rotates counterclockwise.
	Door not sealing tightly.	Adjust the cabinet door position or replace the seal.
	Safety or control thermostats are not operating.	Check the thermostats.
	Bad selector switch.	Check selector switch, see selector switch text.

SERVICE CHART (ALL DRYERS) Continued

SERVICE PROBLEMS	DIAGNOSIS	REMEDY
Dryer does not dry properly.	Low gas pressure in the main line.	Consult the local gas supply company.
	Wrong size orifices.	Consult the local gas supply company.
Pilot burner does not light.	Inoperative pilot valve.	Replace the pilot valve assembly.
	Pilot orifice is clogged.	Check pilot tube orifice — clear the orifice of obstructions.
	Pilot burner filter is clogged.	Replace filter.
	Gas supply line is air locked between the regulator and pilot burner.	Bleed the gas line at the valve connections.
The pilot flame goes out.	Filter is partially clogged. The pilot valve is faulty.	Replace the filter. Repair or replace the pilot valve assembly.
	Gas pressure is low.	Have a local gas company check the pressure.
	Thermo-couple is faulty.	Replace the thermo-couple assembly. (Constant pilot dryers only.)
Main burner does not light.	The main burner orifice is clogged.	Clean the main burner orifice.
	Safety thermostat wiring connections are loose.	Check the wiring or replace the thermostat.
	Temperature control thermostat is inoperative.	Replace the thermostat.
	Main burner air shutter valve is improperly adjusted.	Adjust the air shutter valve on the main burner.
	Gas supply line is air locked between the regulator and the gas solenoid.	Bleed the gas line at valve connection removing air lock.
	An inoperative gas solenoid.	Replace the gas valve solenoid.
	Pilot burner is carbonized.	Clean the pilot burner with a fine emery cloth or a soft wire brush.
	The motor centrifugal switch contacts are bad.	Check the motor, see electrical motors text.
	The gas presure is low.	Have a local gas company check the gas pressure.
	Faulty thermo-couple.	Replace the thermo-couple.
	Gas filters partially clogged.	Replace the filter.
Main burner does not operate properly (burner does not shut off intermittently during the drying cycle or when the door is open).	The motor wiring connections are loose, or the timer is not working.	Check the motor, see electrical motor text. Check the timer, see electrical, timer text.
	Bad door switch.	Replace the door switch.

SERVICE CHART (ALL DRYERS) Continued

SERVICE PROBLEMS	DIAGNOSIS	REMEDY
	Faulty gas valve.	Repair or replace the gas valve.
	Faulty Temperature control thermostat.	Replace the thermostat, see electrical control thermostat text.
Main Burner does not light or operate properly. ALL AUTOMATIC IGNITION MODELS	See text on burners, for all diagnosis and solutions. These are the burner manufactur's recommendations for repair.	
The drum turns, but the heater is not energized.	A broken heating. element.	Remove wires from heater post, check voltage at wires. Test the heater terminal posts for continuity. See electrical element text.
	Inoperative timer.	Check the timer, see electrical timer text.
	Faulty Selector Switch.	Check selector switch, see electrical selector switch text.
	Loose terminal.	Check and tighten all connections.
	Inoperative thermostat.	Check the thermostats both cycling and safety.
	Inoperative motor switch.	Check motor.
	Broken wire in wiring harness.	Check individual wire continuity.
The heater gets hot intermittently.	Natural function of temperature control Thermostat.	None. (This will occur if the timer is set for too long a period.)
Element burns out frequently.	High voltage (245 volts to 265 volts). Part of front felt seal is gone.	Replace the element coil with special high voltage coil. Check the front seal and replace parts as necessary. See section 4, mechanical, front drum seal.

SERVICE CHART (GAS DRYER)

SERVICE PROBLEMS	DIAGNOSIS	REMEDY
Dryer does not dry properly	Low gas pressure in the main line (gas models).	Consult the local gas supply company.
	Wrong size orifices.	Consult the local gas supply company.
Pilot burner does not light.	Inoperative pilot valve.	Replace the pilot valve assembly.
	Pilot orifice is clogged.	Check pilot tube orifice — clear the pilot tube orifice of obstructions.
	Pilot burner filter is clogged.	Replace filter.
	Gas supply line is air locked between the regulator and pilot burner.	Bleed the gas line of the valve connections.
The pilot flame goes out.	Filter is partially clogged. The pilot valve is faulty.	Replace the filter. Repair or replace the pilot valve assembly.
	Gas pressure is low.	Have a local gas company check-the pressure.
	Mercury element or latching mechanism is defective.	Constant pilot dryers only.

SERVICE PROBLEMS	DIAGNOSIS	REMEDY
Main burner does not light.	The main burner orifice is clogged.	Clean the main burner orifice.
	Safety thermostat wiring connections are loose.	Check the wiring or replace the thermostat.
	Temperature control thermostat is inoperative.	Replace the thermostat.
	Main burner air shutter valve is improperly adjusted.	Adjust the air shutter valve on the main burner.
	Gas supply line is air locked between the regulator and the gas valve solenoid.	Bleed the gas line at valve connection removing air lock.
	An inoperative gas valve solenoid.	Replace the gas valve solenoid.
	Pilot burner is carbonized.	Clean the pilot burner with a fine emery cloth or a soft wire brush.
	The motor centrifugal switch contacts are bad.	Check the motor.
	The gas pressure is low.	Have a local gas company check the gas pressure.
	Gas filters partially clogged.	Replace the filter.
Main burner does not operate properly (burner does not shut off intermittently during the dryer drying cycle or when the door is open).	The motor wiring connections are loose, or the timer is not working.	Check the motor. Check the timer.
	Bad door switch.	Replace the door switch.
	Faulty gas valve.	Repair or replace the gas valve.
	Faulty temperature control thermostat.	Replace the thermostat.
Main burner does not light or operate properly. ALL AUTOMATIC IGNITION MODELS.	See gas burner section for all diagnosis and solutions. These are the burner manufacturer's recommendations for repair.	
Dryer does not shut off - Continues to dry - Overdry Auto-dry models only.	Room temperature too cold.	Raise room temperature.

SECTION 2

SERVICE PROCEDURE ELECTRICAL COMPONENTS

Before attempting to service an automatic dryer of any make, the service-man should be equipped with the proper tools. Many of these tools are designed to test the electrical system and components quickly and accurately. Special tools of this type include a test lamp or voltmeter, a continuity tester or ohmmeter and a wattmeter. Proper use of these tools will help make fast, efficient diagnosis and service much easier.

Always use caution when checking any part of the electrical system. Never use an ohmmeter with the machine plugged in. Also, as a safety precaution, always disconnect the electrical power from the dryer before attempting to remove any parts from the machine. For testing purposes, the power cord can again be plugged into a live receptacle after the necessary parts are removed.

All dryers have a wiring diagram attached usually to the back of the cabinet. Study this diagram carefully before proceeding with any electrical checks.

ELECTRIC MODELS

Franklin dryers can be wired to operate on 115 or 230 volts, single phase, 60 Htz. alternating current. All dryers are wired at the factory for 230 volt operation. The dryer operates much more efficiently on the higher voltage. Drying time is shorter when operating the dryer on a 230 volt circuit. Dryers equipped with automatic termination must be operated only on 230 volts.

Wiring

The wiring of a dryer must conform to the local electric code. The dryer should be installed in a well ventilated area, and flammable materials or liquids should never be stored in the dryer area. This holds especially true for gas dryers.

The 230 volt circuit should be fused with a 30 ampere fused switch and come directly from the main switch box. No other appliance or electrical apparatus should be connected to this circuit. The minimum wire size to be used is a number ten.

At least three feet of extra pigtail should be left at the dryer to allow the dryer to be moved for servicing.

CAUTION: The dryer must be grounded for 230 volt operation. Connect the green wire from the ground screw to the N on terminal block. Another alternative is to ground the cabinet to a cold water pipe using the ground screw at left rear of the dryer base.

A terminal block is used to connect the power source to the dryer harness. The harness wires are secured to the terminals by three separate nuts, *Figure 1*.

Three additional nuts and flat, iron washers are furnished for connecting the pigtail to the terminals. Before connecting the dryer to the circuit, be sure the nuts on each terminal are securely tightened. Install the pigtail service cord to the terminals using the washers and nuts supplied.

115 Volt conversion

If the dryer is to be operated on 115 volts, the service cord must be connected as shown in *Figure 1*. Connect one side of the service cord under terminal N and the other side to terminal L1. A jumper wire, at least a number 12 size, must be connected from terminal N to terminal L2 on the terminal block, refer to *Figure 1*.

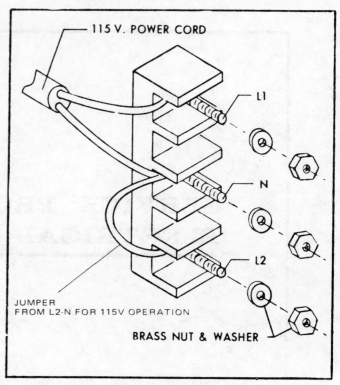

Figure 1 - Terminal Block

If the service cord is a three wire cord with a ground plug, the green wire of the service cord should be connected to the cabinet and NOT to the center terminal marked "N".

CAUTION: For 115 volt operation, dryer must be grounded to a cold water pipe using the ground screw at left rear of the dryer base. Do not connect dryer until jumper wire is in place and ground wire is connected when using the dryer on 115 volts.

ELECTRIC CONNECTION

Gas Models

The dryer is designed for 115 volt operation. Each dryer is equipped with an approved three wire service cord. The plug must be inserted into a compatible receptacle and the receptacle must be properly fused and grounded. A 15 ampere fuse is recommended for use with the gas dryer.

GAS CONNECTION

All models gas dryers

Approved flexible gas lines are available with a gas shut off valve. The flexible line should be a 3/8 tube size. A reducer from 1/2 inch pipe to 3/8 pipe reducer can be used if necessary. Use a pipe sealer on all pipe connections. Flare fittings do not require a sealer. A pressure regulator is standard equipment on all gas models to be used on natural gas.

Do not use a pressure regulator on bottled (L.P.) gas. The regulator is furnished with the tank from the supplier.

Standing Pilot Models

Check the gas shut off valve, *Figure 2.* Be sure it is open. In the open position the handle is parallel to the incoming supply line. The gas line must be bled of air. To do this, push in the red button and hold it in until a small amount of gas escapes through the pilot, *Figure 2.*

CAUTION: Do not attempt to light the pilot until the leaked gas has been cleared away from inside the dryer. Opening the loading door will help in dispersing the escaped gas. All gas connections should be tested for leaks using a soap and water solution or a bottled bubbler solution. Do not use an open flame to test for leaks.

On all gas dryers the gas controls are located behind the access panel. Removing the panel exposes the gas shut off valve, the pilot valve and other functional units and components.

Lighting instructions are as follows:

1. Turn shut off valve to "on" position. Handle should be parallel to the gas supply line.

2. Press red button, *Figure 2.* Hold it in as you place a lighted match over the pilot burner as illustrated in *Figure 2.*

3. Continue to hold button for approximately 30 seconds after a flame is established.

4. Release red button. If flame remains on, dryer is ready for use.

CAUTION: If pilot flame goes out when button is released or during lighting procedure, repeat the whole operation. Wait five minutes before relighting.

5. To shut off gas, return handle to its original position (90°) to the supply line.

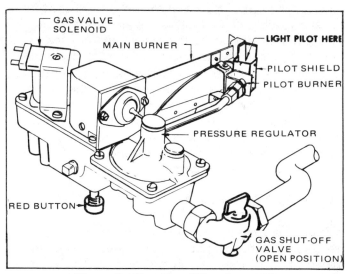

Figure 2 - Standing Pilot Assembly

BURNER ADJUSTMENT

All gas models

The main gas flame is adjusted by opening or closing the air shutter located at the neck of the main burner, *Figure 2A.* Loosen the lock screw and close the air shutter until the flame is almost a solid yellow. Then slowly turn the shutter open until the flame is blue with an occasional orange flash. If the flame is smokey, the shutter is not open enough. Lock the set screw after the shutter is properly adjusted. A most accurate adjustment can be made to the main burner after it has been burning for approximately five minutes. Adjusting instructions apply to all types of gas appliances.

The pressure regulator is preset at the factory to 3.5 water column inches. If further adjustment becomes necessary your local gas company can be called to make these adjustments.

Reference to all electrical components in this section apply to both gas and electric models with the exception of the electric heating element used on electric dryers.

OPERATION

Air Flow

Air is pulled into the drum through the back. It is circulated through the clothes and out the front of the drum. In electric models the fresh air is pulled into the drum as it passes through and around the heat housing. The air then passes through the clothes and out of the drum through the inner panel of the loading door. Finally the air passes through the lint trap, then into the blower and out the exhaust, *Figure 3.* In gas models, air is pulled from the combustion chamber, where it is heated, and then into the heat housing and on into the drum, *Figure 4.* By having the blower pull the air from the drum instead of pushing it into the drum, a negative pressure or a slight vacuum is established. This negative pressure results in faster evaporation of moisture at a given temperature.

Automatic Termination

Figure 5 illustrates the automatic termination cycle, showing the approximate temperatures. It also indicated the time the heat source and timer motor are in the circuit for one complete cycle.

During the initial heating period, the single pole, double throw switch in the control thermostat completes a circuit to the heater leaving the timer motor circuit "open".

Satisfying the thermostat will open the heater circuit and complete a circuit to the timer motor.

DRYERS WITH AUTOMATIC TERMINATION MUST BE OPERATED ON 230 VOLTS.

The switch in the control thermostat prevents a closed circuit to the heater and timer motor at the same time, *Figure 6.*

TIMED DRYING

All models have a timed drying cycle. Some models feature a selector switch which allows selection of "high", "medium" or "low" heat.

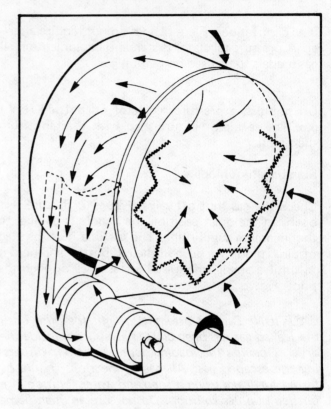

Figure 3 - Air Flow Diagram - Electric Models

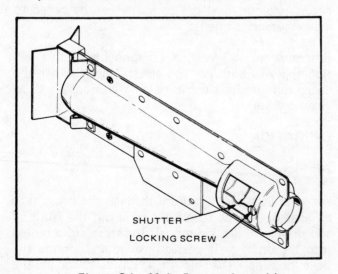

SHUTTER

LOCKING SCREW

Figure 2A - Main Burner Assembly

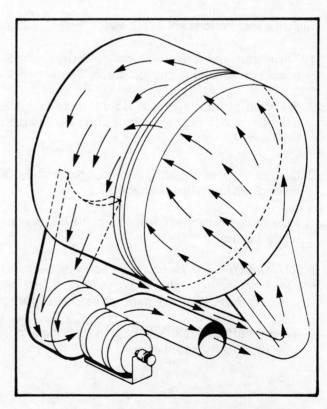

Figure 4 - Air Flow Diagram - Gas Models

NOTE: The heater is out of the circuit for the last five minutes providing a cool down period for both the timed drying and automatic termination cycles.

THERMOSTATS, LATER MODELS

Each model dryer is equipped with a control or cycling thermostat which is located in the exhaust duct, adjacent to the blower, also a safety thermostat or high limit thermostat located in the heat housing.

On models designed for multiple heat selection, the cycling thermostat incorporates a built in heater. For some heat selections, the selector switch places a resistor in series with the thermostat heater, *Figure 7.*

TIMER, LATER MODELS

The timer is an "on-off" switch operated by a motor driven cam. Turning timer "on" completes an electrical circuit to other controls and components and then starts dryer operation.

All timers used are functionally the same but differ in number of cycles, length of cycles, length of shafts, terminal locations, terminal markings and appearance. These differences generally result from the styles of manufacture. Four distinct brands are used. These are illustrated in *Figure 8* for convenience of identification.

All timer heat cycles have a pre-set cool down period at the end of the drying cycle, (varies five to ten minutes depending on cycle and model) with the exception of the auto permanent press cycle which has a seventeen minute cool down period of tumbling without heat.

The timers, regardless of brand, utilize three to four contact points. The terminals emanating from these are considered the power terminals. The markings and utilization of each are as follows:

Markings	Use
B or L1	Hot terminal as it is connected to incoming power line.
C or M	Drive motor and light circuits.
A or H	Ignition circuits.
TM (Some Models)	Timer motor circuit.
B	End of cycle signal.

Remaining terminals, regardless of markings are common terminal points. To gain access to the timer for checking circuits or replacement, remove the control panel and backguard rear shield or control mounting bracket.

Timers of different makes are used interchangeably. As a result, the replacement you receive may look different from the one originally supplied on the dryer. When connecting the wires to the replacement, make sure they are attached to the proper terminals, i.e., wires from A of the original to A of replacement, B of the original to B of the replacement, C of the original to C of replacement.

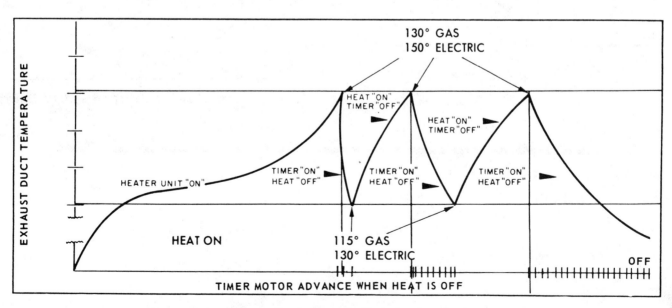

Figure 5 - Automatic Termination Cycle

When connecting the wires to the common junction points, remember that the connections fall into two groups, one group consisting of mostly yellow wires and the other mostly white wires. These wires are to be transferred one at a time. Since there are more wires to the white group it will be necessary.

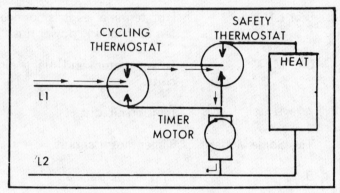

Figure 6 - Automatic Termination Circuit

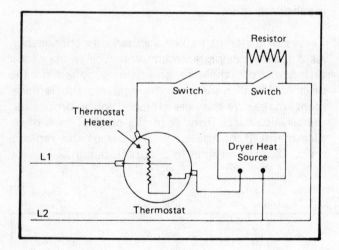

Figure 7 - Control Thermostat Circuit

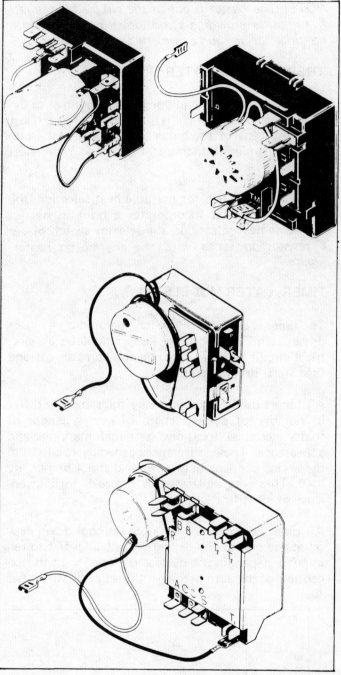

Figure 8 - Illustrations of Various Brands of Timers

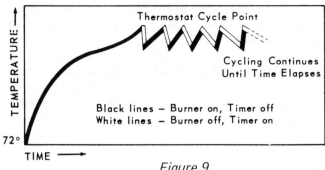

Figure 9

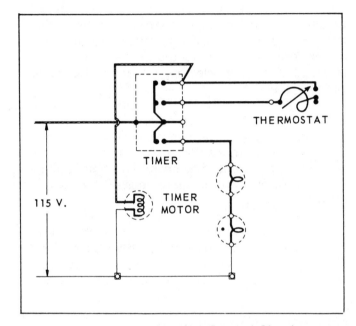

Figure 10 - Automatic Dry Control Circuitry

Figure 11 - Safety Thermostat

AUTOMATIC DRY CONTROLS - Some Models

The automatic dry control provides the user with a choice of two basic drying cycles. They are the TIME CYCLE and the AUTO CYCLE.

The TIME CYCLE is used for drying special garments. When placed at this setting the dryer operates in the same manner as dryers without the dryness control. The temperature control (thermostat) is set for the type of load and governs the temperature in the drying chamber, and the timer is then set for the desired drying time. When the selected time has elapsed, the dryer will shut off. A no-heat cool down period is included at the end of the drying time.

The AUTO CYCLE is used for all average to large loads. It provides an automatic shut-off of the dryer according to the setting selected by the user. The temperature control (thermostat) is set at the desired temperature setting. The timer is set to a recommended setting (on control panel) for type of finished load desired. When using the A or D settings, turn the knob clockwise to B or E and then back to A or D. It is possible the heat circuit may not be completed and as a result the machine never stops running. The operator instructions also show suggested settings. Drying is accomplished through the relationship of the timer and thermostat. The timer will not advance until the thermostat has cycled (does not call for heat), *Figure 9*. This is accomplished by the use of a S.P.D.T. thermostat in the circuit. When the thermostat has cycled (does not call for heat), it completes the timer motor circuit, *Figure 10*. As the thermostat again calls for heat, the timer motor circuit is opened. This on and off cycling continues for a predetermined length of time for the setting selected. After the heating cycle is completed, a no-heat or cooling period is provided. Terminal TM on the timer is then completed and allows the timer to drive to the off position.

SAFETY THERMOSTAT, LATER MODELS

NOTE: The door switch is not considered as a basic component of the automatic dry control system.

The safety thermostat, a single throw switch, mounts on the drum heat housing of both the gas and electric models. If an air flow block raises the dryer temperature, a bi-metal disc in the thermostat opens the circuit to the heat source, and allows the dryer to cool. On gas and electric dryers, the safety thermostat opens at 210° and closes at 130°, *Figure 11*.

To Check the Thermostat:

1.   Remove the dryer back panel.

2. To check for stuck contacts in the thermostat:
   A. Start the dryer and run it on high heat. With the exhaust duct completely blocked, this thermostat must open within three minutes, (dryer empty).

3. To check for an open thermostat (power source disconnected):
   A. Remove the wiring from the safety thermostat terminals.
   B. Test the thermostat for continuity with a test light.
      The light shows continuity through a good thermostat at room temperature.

   High Heat — No heater or resistor in circuit.
   Med. Heat — Heater and resistor in circuit.
   Low Heat — Heater only in circuit.

Drum Temperatures

Resistors of different values, combinations of resistors and/or a heater in the control thermostat, are used to provide varying drum temperatures.

The circuitry involving the thermostat heater and the resistor vary with each heat selection to produce the different temperatures.

High Temperature — circuits through the external resistor and the thermostat heater are open. Only the temperature from the heat source effects the control thermostat.

Medium temperature — circuits close through the resistor and the thermostat heater. Heat generated by the thermostat heater supplements the dryer heat, causing the thermostat switch to open sooner, resulting in lower drum temperatures. The thermostat heater cycles on and off with the heat source, on gas models only. The thermostat heater is on constantly in the electric models.

Low temperature — no resistors are in the circuit. The thermostat heater cycles on and off with the dryer heat source, on gas models only. The thermostat heater is on constantly in electric models. Without the resistor in the circuit, more current flows through the thermostat heater, resulting in lower drum temperatures.

Resistors

The resistor, located in the circuit between the push-button or rotary selector switch and the thermostat heater, reduces current through the heater. Depending on the program switch setting, current through the thermostat heater varies. This current variation causes the thermostat heater to generate varying amounts of heat.

The resistors are checked with an ohm meter. The resistor values are marked on the schematic wiring diagram. A bad resistor will give improper drying temperatures. The resistor assemblies are mounted on the back of the push-button or rotary selector switch.

TIMER ASSEMBLY

The timer has a clock type motor and electrical switches. Moving the timer control knob away from the OFF position completes a circuit to the dryer motor. Another circuit to the heat source is completed through the selector switch and the motor centrifugal switch.

The timer may have an automatic termination and time-dry cycle, or just a time-dry cycle. Refer to the timer dial to determine the length of the drying time of the time-dry cycle. The operation of an automatic termination cycle is approximately twenty-one minutes.

The timer may be checked with either an ohm-meter or a test light, which is fused with an 8 amp fuse. The timer has either two or three sets of contacts which control dryer heat, the drum motor, and the timer motor.

SELECTOR SWITCH - Some Models

The six button, push-button selector switch controls drum temperatures for the cycle selected. Refer to the fabric selection guide, on backguard trim plate, *Figure 12*.

Fabric Selection Guide					
AUTO CYCLE			**TIMED CYCLE**		
1. Cotton & Linen			4. High		
2. Perma-Press			5. Low		
3. Damp Dry			6. Air Fluff		
Push Button No.	1	2	3	4	5	6
Contacts Closed	A to E, T to J, F to B	A to E, J to P, F to B	A to E, L to J, F to B	A to E, S to J, G to B	A to E, K to J, G to B	G to B

Figure 12 - Selector Switch

Remove backguard backplate for access to selector switches. For replacement or to make a continuity check, on all models, refer to *Figure 13*, continuity check and *Figure 15*, selector switch.

NOTE: When making a continuity check, be sure to remove all harness wires and resistors from switch terminals.

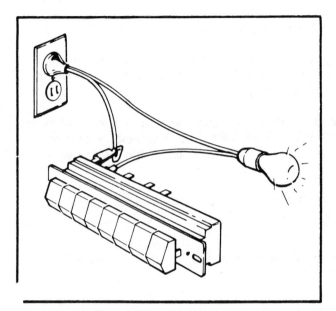

Figure 13 - Continuity Check

ROTARY SELECTOR SWITCH - Some Models

The four position, rotary selector switch, allows selection of high, medium and low temperatures and air fluff.

NOTE: Number one position on all rotary selector switches is when the switch knob is rotated counter-clockwise, as far as possible.

Refer to *Figure 14* when making a continuity check.

POSITION	CONTACTS
1. CCW Limit	No Contacts Made
2.	A to E, J to L
3.	A to E, J to K
4. CW Limit	J to K

Figure 14 - Rotary Selector Switch

Another type of four position, rotary selector switch allows selection of air fluff, also three temperatures, low, medium and high, *Figure 15*.

NOTE: When air fluff is desired, timed cycle drying must be used.

Contacts Closed	Switch Position
None	1. Air Fluff
A-E, J-L	2. Low Heat
A-E, J-K	3. Medium Heat
A-E	4. High Heat

Figure 15 - Selector Switch

TWO POSITION ROTARY SELECTOR SWITCH

The two position rotary selector switch allows selection of high temperature only and air fluff.

Refer to *Figure 16*, when making a continuity check.

POSITION	CONTACTS
1. CCW Limit	Closed
2. CW Limit	Open

Figure 16 - Two Position Switch

FLUORESCENT LAMP

The 15 watt lamp is mounted inside the backguard.

A momentary contact switch is mounted at the right side of timer dial. Press switch button in. When lamp glows, release button.

Remove trim plate for access to lamp or switch.

Ballast

The ballast mounted inside the backguard is required to provide sufficient voltage for starting the lamp and limiting the flow of current to the lamp after it is illuminated.

Without a ballast, the lamp will be destroyed instantly if 115 volts is connected direct to the lamp.

Remove back backguard backplate for access to ballast.

Checking Procedure

1. Be sure lamp pins are tight and make good contact in sockets. Check at timer terminals to be sure current is available to lamp.

2. Check lamp filaments with an ohm meter or bulb checker only. There should be continuity between the two pins at each end of lamp.

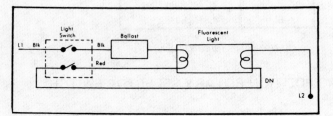

Figure 17 - Fluorescent Schematic

3. There should be continuity through the ballast.

4. With the switch in off position, press the button in to start position. This closes both sets of contacts for continuity through each pair of (black and red) wires. Releasing the button, so switch returns to on position, opens the momentary closed contacts to show continuity only through the black pair of wires, *Figure 17.*

BUZZER

The buzzer, located in the backguard, signals when the timer opens the circuit to motor at the end of each cycle.

The buzzer signals each time the loading door is opened.

Remove backguard backplate, for access to buzzer.

DOOR SWITCH

The switch is located at upper right hand corner of recess for loading door.

The switch is of the single-pole, double-throw type, and is actuated by contact of the door. When door is opened, the switch automatically stops all dryer operation. When door is closed, the dryer operation is resumed.

The switch is mounted into front of cabinet wrapper with a hex nut.

The door switch can be checked or replaced by raising the top panel.

On some models, the interior drum lamp is also controlled by the door switch to illuminate the drum interior when door is opened.

DRUM AND OZONE LIGHT - Some Models

The 40 watt, heavy duty appliance, incandescent lamp lights the drum interior whenever the loading door is opened.

The blue germicidal ozone light bulb is wired in series with the drum light. The incandescent lamp acts as a ballast for the ozone light which is one whenever the timer motor operates.

CAUTION: Do not test the bulb directly with 115 volts.

Should the drum light fail, the ozone light will not function.

Remove cabinet rear panel for access to the lamps, *Figure 19.*

THERMOSTATS

Control Thermostat

The control thermostat is located in the outlet duct which connects to the outlet of blower housing, *Figure 18.*

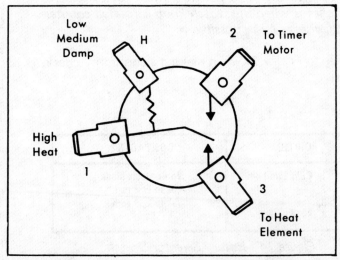

Figure 18 - Four Terminal Thermostat

The thermostat switch is actuated by a bi-metal disc.

Three types of control thermostats are used on the different models:

1. For multiple heat dryers with automatic termination, the thermostat contains a heater and a single-pole, double-throw switch. See *Figure 18.*

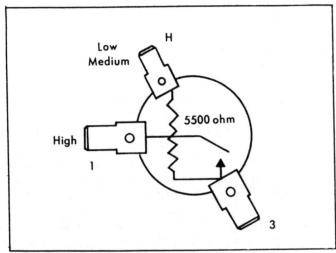

Figure 20 - Three Terminal Thermostat

2. For multiple heat dryers, the thermostat contains a heater and a single-pole, single-throw switch, *Figure 20.*

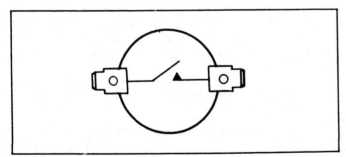

Figure 21 - Two Terminal Thermostat

3. For single temperature dryers, the thermostat contains a single-pole, single-throw switch, *Figure 21.*

Checking The Control Thermostat

Remove harness wires from the thermostat. Determine the interior wiring of the thermostat by referring to *Figures 19, 20, 21,* or the schematic wiring diagram. Use an ohm meter or a light bulb, as illustrated in *Figure 12,* to check the thermostats.

1. On a four terminal thermostat, there should be continuity through the heat contacts and through the thermostat heater, but not through the timer motor contacts at room temperature.

2. On a three terminal thermostat, there should be a circuit through the heat contact and through the thermostat heater at room temperature.

3. On a two terminal thermostat, there should be continuity at room temperature.

To check for stuck contacts, repeat these tests and apply heat to the thermostat (lighted match) to determine whether or not the contacts open and then close when it cools.

Safety Thermostat

The safety thermostat, single-pole, single-throw switch, wired in series with the control thermostat

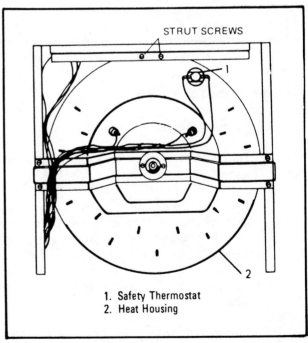

1. Safety Thermostat
2. Heat Housing

Figure 19 - Safety Thermostat - Electric Models

and heat source, mounts in the heat housing on both the electric and gas dryers. See *Figures 19* and *18A* for mounting location of thermostats.

Should the control thermostat fail or an air block occur in the vent system which would cause excessive drum temperatures, the safety thermostat opens the circuit to the heat source and prevents over heating.

The safety thermostat opens at 210° and closes at 130° on both the electric and gas dryers.

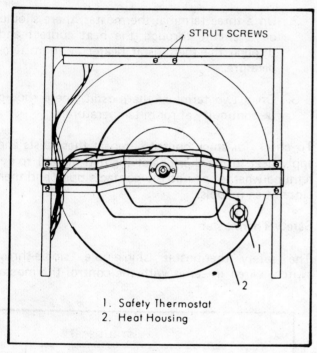

1. Safety Thermostat
2. Heat Housing

Figure 18A - Safety Thermostat - Gas Models

Checking the Thermostat

1. Remove the cabinet rear panel.

2. To check for stuck contacts in the thermostat:

   Start the dryer and operate it on high heat. With the exhaust duct completely blocked, the thermostat should open within about three minutes. (Dryer empty).

3. To check for open contacts in the thermostat:

   Disconnect power, remove the wiring from the thermostat terminals.

Remove the wiring from the thermostat terminals.

Make a continuity check. The switch contacts will be closed in a good thermostat, at room temperature.

ELECTRIC MOTOR

The ¼ hp, 115 volt, 60 HTZ, single phase, 1725 rpm, double shaft motor is mounted in a cradle. The cradle is fastened to a base with a bolt at front, and at back with a tab which slips into a slot in the base.

The blower impeller, which has an insert with left hand thread, is mounted on the end of the motor shaft and the drum drive pulley is mounted at the opposite switch end of the shaft with a set-screw.

NOTE: General Electric, Emerson or Westinghouse motors may be used and are completely interchangeable. Each different motor requires a different centrifugal switch. Refer to the parts catalog for the correct switch.

When the motor is energize, the run (main) winding is in the circuit at all times. The start winding is in the circuit only during the starting period. The start winding determines the direction of motor operation and gives torque for starting.

The motor consists mainly of a run (main) winding, start winding, centrifugal switch, rotor and shaft.

The rotor (revolving portion of the motor) is attached to the shaft. The moving portion of the centrifugal switch is also attached to the shaft. The replaceable (terminal block) stationary portion of the

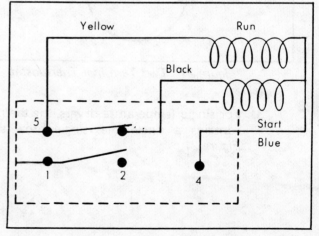

Figure 22 - Motor Schematic

centrifugal switch controls the circuit to the start winding and heat source. When the motor is idle, the switch to start winding is closed and the switch to heat source is open.

When current flows to the motor, the rotor starts to revolve and as it reaches operating speed, the moving portion of the centrifugal switch actuates the stationary portion of switch. This opens the circuit to the start winding and closes the circuit to heat source (terminals 1 and 2). See *Figure 22* which shows a schematic of the motor when it is idle.

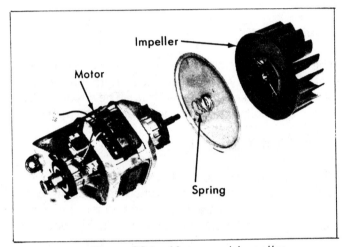

Figure 22A - Motor and Impeller

Checking the Motor, *Figure 22A*

1. Remove rear cabinet panel.

2. Remove harness wires from motor.

3. Operate motor by connecting a properly fused service cord to terminals 4 and 5. The motor should start and run.

4. When motor will not run, remove terminal block which is fastened to the motor with two screws.

5. Make a continuity check between the black and yellow motor wires or wires which connect to terminals 5 and 4.

    No continuity shows an open in one of the windings.

6. With continuity on the above check, remove the wires at inside of switch. Then, check from terminal 5 to inside center black wire

terminal or to the black wire which connects to terminal 6 with the switch lever pressed in start position.

No continuity shows the switch is inoperative.

7. When the motor runs and the problem is NO HEAT, check at terminals 1 and 2. The switch lever must be out (run position).

    No continuity shows the switch is inoperative.

CAUTION: The motor must be removed from the dryer base to check for a ground.

Overload Protector

The motor overload protector, imbeded in the windings of the motor, is not replaceable.

Should the protector fail, the motor must be replaced.

The specific purpose of the overload protector is protection to prevent a motor burn-out, in case an electrical or mechanical overload occurs.

HEATING ELEMENT

ELECTRIC DRYERS, All Models

The electric heating element rating is 5000 watts at 230 volts, and 1250 watts at 115 volts. Where high voltage is commonly found, install the 256 volt element.

NOTE: Automatic termination models will not operate properly on 115 volts; the cycle will not terminate.

Checking The Element

1. Remove rear cabinet panel. (Power source disconnected).

2. Remove harness wires from element terminals and make a continuity check at terminals.

3. Make a continuity check from each terminal to heat housing to determine whether or not a ground exists.

Element Replacement

1. Refer to drum removal text for removal of heat housing.

2. Remove old element from terminals, then remove all insulators and clips from heat housing.

3. Locate all insulators and clips on new pre-stretched element (60 inches long).

4. Connect element to both terminals.

5. Install NUMBERED insulators and clips in heat housing, as shown in *Figure 23*.

6. Carefully divide element equally around heat housing, then install balance of insulators and clips.

   CAUTION: When connecting the element to terminals, always wrap wire clockwise around terminal. Install washer and nut, tighten nut securely to be sure of a good electrical connection.

WIRING HARNESS

The wiring harness, not available as a complete component, can be replaced as individual wires. Refer to the wiring diagram for the correct size and type of wire at each location.

To remove the motor and impeller fan:

1. Release the spring holding the idler arm.

2. Disconnect the harness at the motor.

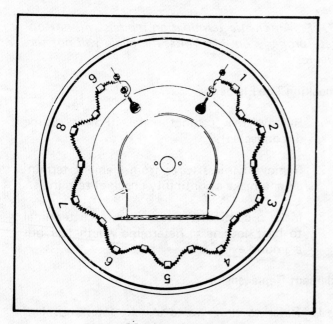

Figure 23 - Heating Element

3. Remove the screw holding the motor mount.

4. Pull up and out on the motor.

5. Rotate the assembly to remove it from the dryer.

To remove impeller fan:

1. Hold motor shaft with a vise grip and turn the impeller clockwise.

2. The cover plate and spring slide off the shaft.

LIGHTS

Interior Lights

The incandescent bulb used inside the dryer is a 40 watt, heavy duty appliance bulb. It mounts at the rear of the dryer behind the drum. The circuit closes to the lightbulb whenever the loading door opens.

Ozone Light, Figure 23B

This is a blue germicidal bulb, wired in series with the interior incandescent bulb. The incandescent acts as a ballast for the ozone light. When the incandescent bulb fails the ozone light will not operate. The ozone light is on whenever the timer motor operates.

Fluorescent Console Light

The bulb is 15 watts. It is locked in place so that it cannot shake loose from vibration.

Fluorescent Light Switch

A momentary contact light switch mounts either to the top of the console or to the face of the trim plate.

Ballast For Fluorescent Light

This component mounts inside the control housing. The ballast does two things.

1. It increases the voltage necessary for starting the light.

2. While the light operates, the ballast stabilizes the flow of current through the tube.

Checking The Fluorescent Light

If the bulb does not light, *Figure 23A.*

1. Check the continuity of the bulb filaments by using either an ohmmeter or a light bulb checker. You should have continuity between the two pins on each end of the bulb.

2. You should have continuity through the ballast.

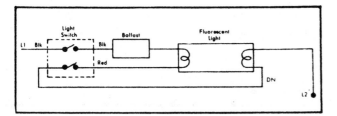

Figure 23A - Schematic for Fluorescent

3. Check the light switch. You should have continuity through the two black wires with the switch in the on position. When the switch is turned to the on position and the button is held in, you should have a circuit through the two red wires.

4. Check each item separately. If all components check good, then check the wiring on the timer to see if the light is receiving voltage.

5. This diagram of a fluorescent light is very much like the diagram found on all light equipped models. If you follow this step-by-step of checking, fluorescent light repairs should be easy.

THE END OF CYCLE SIGNAL

The end of cycle signal is a non-adjustable buzzer controlled by timer cam BD. The buzzer operates about a minute at the end of each cycle.

WIRING HARNESS

The wiring harness is a separate component which may be replaced if necessary. Normally, if you have a bad wire in a harness there will be visible damage to the outside of the wire. Usually you can repair or replace the bad wire individually.

A 40 watt ballast-type flood lamp is located at the top on the right of the drum case wrapper, *Figure 23B.* It provides light in the drum when the dryer is operating or when the dryer is stopped and the loading door is opened. Next to it is the germicidal or ozone lamp. These two lamps are wired in series, the 40 watt lamp acting as a ballast for the ozone lamp. When either of the lamps burn out, the other will not light. To check the lamps, remove the 40 watt lamp and test it in a light socket or with a test cord. If it lights the ozone lamp should be replaced. NEVER TEST THE OZONE LAMP WITH 115 VOLT HOUSE CURRENT.

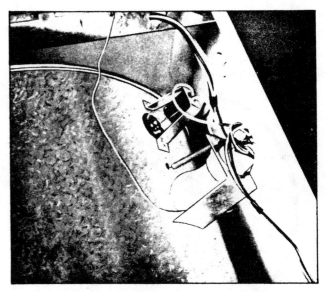

Figure 23B

POWER SUPPLY

EARLY MODELS

For 115 volt installations, a 20 amp circuit is required. On all electric dryers the drying time is considerably longer if the 115 volt circuit is used to supply the heater element. The instructions for circuit hookup should always be consulted when starting to work on a dryer. See *Figure 24*.

Needless to add, all code requirements, national and local, are to be followed when making installations or checking for trouble.

Gas dryers operate on only a 120 volt circuit.

Connecting The Electrical Supply
On some of the models, the backguard shield can be removed to make the connections to the power supply. On other models, the top assembly will have to be removed. The removal instructions are covered elsewhere in the manual. Proper connections are shown in *Figure 25,* for models with three heater terminals. *Figure 26* shows the correct terminal connections for models with two heater terminals.

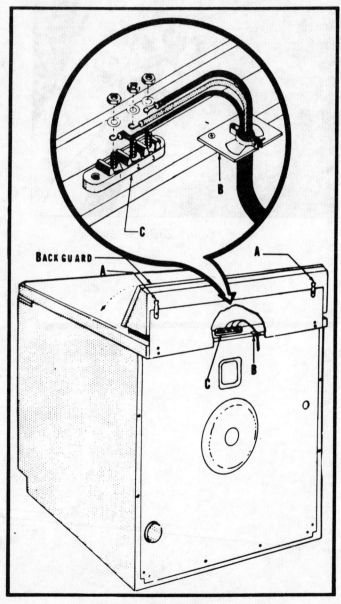

Figure 24

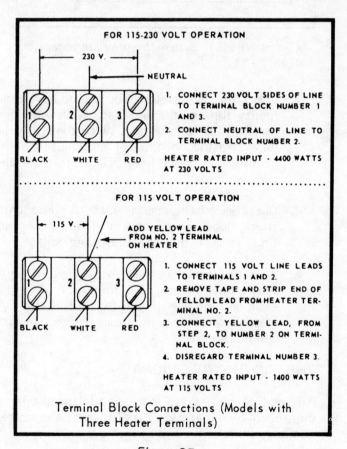

FOR 115-230 VOLT OPERATION

230 V.

NEUTRAL

BLACK   WHITE   RED

1. CONNECT 230 VOLT SIDES OF LINE TO TERMINAL BLOCK NUMBER 1 AND 3.
2. CONNECT NEUTRAL OF LINE TO TERMINAL BLOCK NUMBER 2.

HEATER RATED INPUT - 4400 WATTS AT 230 VOLTS

FOR 115 VOLT OPERATION

115 V.

ADD YELLOW LEAD FROM NO. 2 TERMINAL ON HEATER

BLACK   WHITE   RED

1. CONNECT 115 VOLT LINE LEADS TO TERMINALS 1 AND 2.
2. REMOVE TAPE AND STRIP END OF YELLOW LEAD FROM HEATER TERMINAL NO. 2.
3. CONNECT YELLOW LEAD, FROM STEP 2, TO NUMBER 2 ON TERMINAL BLOCK.
4. DISREGARD TERMINAL NUMBER 3.

HEATER RATED INPUT - 1400 WATTS AT 115 VOLTS

Terminal Block Connections (Models with Three Heater Terminals)

Figure 25

CONVERTING TO 115 VOLT

If the model to be converted from 230 V supply to 115 V supply uses the Sensitron system, the electrical polarity must be observed.

The service entrance wire should be a number 12 wire and the system must be protected by a 20 ampere fuse.

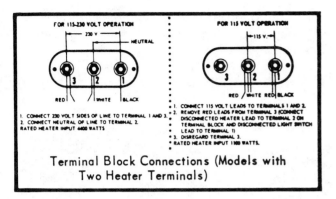

Figure 26

To check the polarity, use a Neon test light to find the "hot" side of the wall receptacle. Mark this plainly. After marking the receptacle, plug the cord in and check the polarity by turning the Sensitron to the "OFF" position. Then, by tilting the panel backguard backward, a probe of the Neon test light can be placed on the contact "L1" at the Sensitron while the second probe can be touched to the ground wire. The Neon light should light. If it doesn't, remove contact from "L1" and place it on the "White" common junction at the Sensitron. If the Neon light now lights, polarity at the Sensitron is reversed. To correct this trouble, reverse the service cord wires at the terminal block.

Of course the dryer must be properly grounded at the service entrance as well as having the correct polarity. Always connect the ground wire from the machine cabinet to a COLD water pipe to effect a proper ground.

NEVER GROUND A DRYER TO A GAS PIPE.

SWITCHES

DOOR SWITCH
The door switch, *Figure 27,* controls both the lights and the stopping of the motor when the door is opened. Some have soldered connections and the others have quick-disconnect terminals. Replacement switches will necessitate clipping the wires from the old one and fitting them with solderless terminals.

When switches become defective, replace them for complete satisfaction. Repairing switches is not recommended.

Replacing the switch can be done by removing the

knurled nut and pushing the switch inside of the cabinet. Pull it up into a workable position and carefully transfer the wires, one at a time, to insure the proper positioning of the wires. Align the two switches using the keyway slot on the switch shank as a guide. To reassemble, reverse the process and make sure that the key of the cabinet front fits the keyway of the switch. By moving the hexagon nut, the end of the plunger can be set at the correct distance from the cabinet front (5/8"). Be sure to tighten the knurled nut securely.

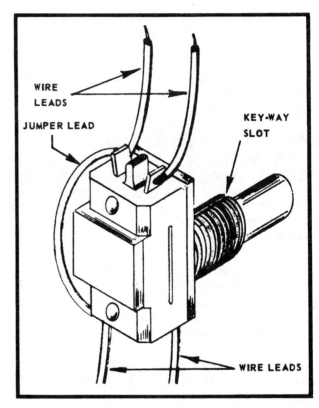

Figure 27

LINT SCREEN SWITCH
The lint screen switch, *Figure 28,* breaks the circuit when ever the lint screen is left out or not fully pushed into place. It is mounted on the foot release assembly and, in most cases, behind the left side of the toe rail.

To replace this switch, remove the lint screen and detach the foot pedal from the foot release assembly by moving attaching screws. Then remove the screws holding the foot release assembly to the toe rail so that the toe rail can be removed completely.

The switch can be removed by grasping it behind

the foot release bracket at the top and the bottom, depressing the clips and pushing the switch forward. It will then be possible to pull the switch.

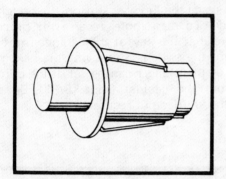

Figure 28

There are quick disconnect terminals that allow the wires to be disconnected. Reversing the procedure will reinstall the switch.

LIGHTS — ULTRA-VIOLET LAMP
A mercury vapor lamp furnishes ultra-violet rays for the germicidal action which destroys germs, fungus and molds. There is a predetermined quantity of ozone released, as well, which also adds to the bacterial destruction and helps to give a clean sweet smell to the clothes.

The bulb is located above the loading port and has a perforated shield intended to guard against bulb breakage and protect the operator's eyes from the direct rays of the lamp. "Sun-E-Day" is imprinted on the shield and identifies the location.

The ultra-violet lamp is a 4 watt bulb connected in series with the interior floodlight. It should NEVER be tested with 115 volts.

If the ultra-violet lamp and the floodlight do not light, remove the flood light and check it with 115 volts. If it proves good, replace the ultra-violet lamp.

INTERIOR FLOODLIGHT
To get at the flood light, the top of the dryer must be removed (explained elsewhere) and the sheet metal screw that anchors the left side of the floodlight bracket removed. The floodlight is located in the top front of the drum insulation area, *Figure 29.* A 40 watt appliance type bulb is used. After the bulb has been replaced, the above steps can be reversed to reassemble.

Since the floodlight is wired in series with the ultra-violet lamp, it acts as a ballast to reduce the voltage to the ultra-violet lamp.

If either of these bulbs is burned out, neither one will light.

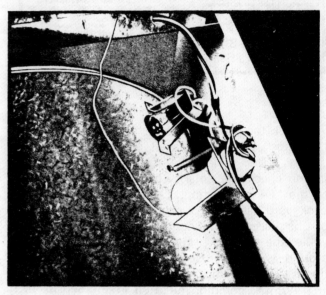

Figure 29

CONTROL PANEL LIGHTS
The majority of control panels have lights that can be serviced through tipping the control panels either forward or backward after loosening the set screws and removing the knobs. In some models, the backguards have brackets that hold the backguard, In others, the backguard is attached to the cabinet frame, *Figure 30, 34.* In each case the bulbs can be serviced after tilting backguard and lifting retaining clips out of the tab slots that help to hold backguard in place.

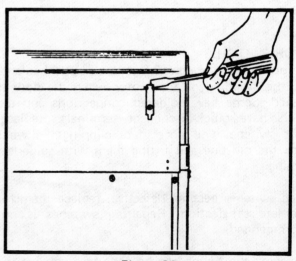

Figure 30

Figure 31

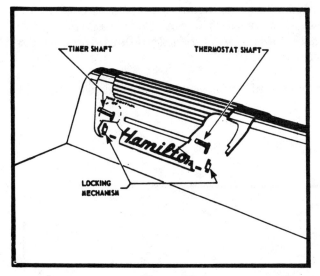

Figure 32

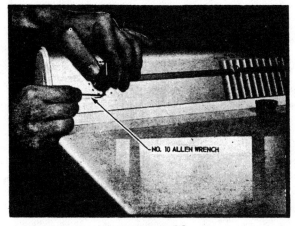

Figure 33

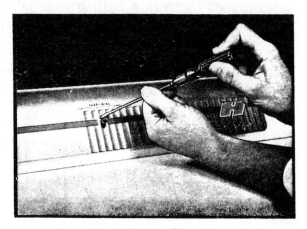

Figure 34

INDICATOR LIGHTS

Special lights on certain models are used to signal specific conditons in the dryer. Such a light is the Lint-Minder indicator, connected in parallel, which lights up whenever the circuit is complete between the drum protectomatic switch, the timer, the door switch and the lint screen switch. If any of these controls open the circuit the red light will go out and the operator will know that something is not right. The most common trouble is forgetting to fully close the lint screen completely, or leaving it out entirely.

Another special light used on some of the top of the line models is the Heat-Miser. This light glows whenever the dryer heater is off. If the operator uses the Fabri-Set thermostat at the AIR position, the light will remain on until the position of the Fabri-Set thermostat is changed. At any of the other settings of the thermostat, the light will come on only when the Fabri-Set thermostat breaks the circuit to the heating element. It will remain on until the heater circuit is again reenergized.

Both of these special indicator lights are removedable by detaching the indicator light bracket from the backguard after removing the control panel. When reinstalling the indicator light bracket, be sure that the black fiber shield is attached to the metal tab of the bracket. The purpose of the shield is to keep stray light from leaking from one indicator to another. This should be checked if BOTH lights seem to stay on. Refer to *Figures 31, 32 and 33*.

DISCONNECT SERVICE CORD FROM THE
WALL RECEPTACLE BEFORE PERFORMING
ANY SERVICE ON THE DRYER.

Note the location of all spacers, cushions, speed nuts
or lock washers during dissambly.

Replace these parts in their proper position during
reassembly.

CABINET ASSEMBLY LATER MODELS

Top Panel

The backguard assembly is mounted on and secured
to the top panel with screws which extend up from
the back underside of top.

The top panel is held in place with two screws at the
back, which extend up through the top flange of
rear panel into the lower flange of top, and two
spring clips at the front. The clips lock into the
front top flange of cabinet.

Two rubber bumpers are located at each side on top
flange of cabinet, between top panel and cabinet,
Figure 35.

During reinstallation of top, carefully arrange all
harness wires to prevent them from contacting the
drum assembly.

Backguard, *Figure 36.*

The backguard houses the timer, selector switch,
buzzer and lamp components. All components in
the backguard are accessible for service by removal
of the trim plate (dial panel) and/or the backplate.

Cabinet Wrapper

The cabinet wrapper is fastened to the base with
three screws at each side, one screw at each side in
the front under the access panel and two screws
through the bottom or rear panel.

BACKGUARD ACCESS, *Figure 36*

Controls housed within the backguard are timer,
thermostat control panel lights, end of cycle signal,
(depending on model involved).

For rear access to these components proceed as
follows:

1. Lay drop cloth over cabinet top so that cloth
   extends to backguard.

2. Loosen set screws and remove control knobs.

3. Remove three screws at bottom front of backguard, securing control panel and end caps to
   backguard. Grasp the bottom edge of the control panel, rotate upward until the top flange
   can be pulled forward out of the backguard
   and remove.

4. The individual controls can now be removed
   from dryer backguard.

*NOTE: When access to rear of controls by means of
removing the backguard rear shield is difficult, due
to dryer relationship of walls or cabinets, access can
be accomplished from the front by removing the
backguard control mounting bracket. It is secured to
the backguard with screws. Once the screws are removed, simply pivot it forward out of unit, as illustrated in Figure 36.*

To reassemble, reverse above procedures.

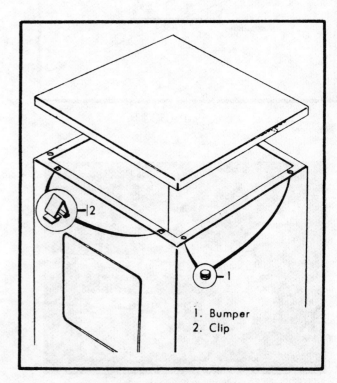

1. Bumper
2. Clip

Figure 35 - Cabinet Assembly

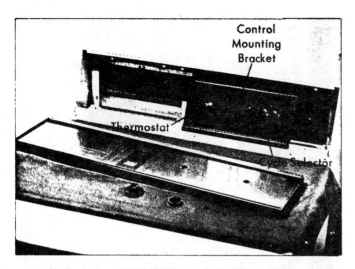

Figure 36 - Servicing Backguard Controls

TOP PANEL

1. Remove the top panel to:
   A. Replace or repair
      1. Door switch,
      2. Door hinges,
      3. Rear drum bearing.

   B. Inspect drive belt

   Two screws (Item 1, *Figure 37* hold the top at the rear, and two clips (Item 2, *Figure 37* hold the panel in the front. To remove the top, take out the screws and pull the top forward to clear the clips. When replacing the top, see that the wiring harness is clear and that the drum will not rub it.

2. Loading Door, *Figure 38*.

   The loading door parts can be replaced separately. To remove the door, take out the four screws, (*Figure 38*, holding the door to the hinges

   A. Door Handle
      The door handle can be removed without removing the door. Only loosen, do not remove the screws. If screws are removed, inside spacers will fall out.

   B. Inner Panel Strip Gasket
      This gasket is a seal between the lint filter and the drum support ring.

C. Door strike
   The door strike is press fit into a square hole. The strike expands to hold itself in place.

LOADING DOOR

All door parts are replaced separately, *Figure 38.*

Each hinge is fastened to the cabinet by flat head bolts with nuts and shakeproof washers located inside of cabinet. Top panel must be raised to gain access to nuts and washers.

Each hinge is fastened to edge of door with phillips head screws which extend into speednuts.

The door is removable for access to all components, *Figure 38.*

The strip filler gasket, located on inner door panel under air outlet, must be in place to prevent air leakage between line trap and support ring.

Door Latch

The latch is held in cabinet opening with a spring clip which is a part of the assembly.

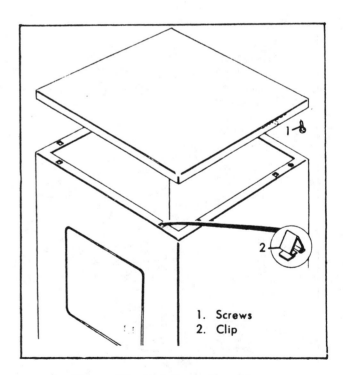

1. Screws
2. Clip

Figure 37 - Top Panel Mounting

The latch may be removed and reused if it is carefully pried from opening. Be sure to protect painted surfaces from mars or scratches. Pressure on latch from inside of cabinet is recommended during removal. Install latch by pushing it in opening, *Figure 38.*

LOADING DOOR GASKET

To remove gasket, lift the lip and loosen all screws holding the support ring in place. Remove the gasket from support ring assembly and slip the new one in place, *Figure 39*

On reassembly, when the support ring is removed, be sure the front throat of drum is in place on the two bearings at top of support ring and the spring loaded hold down bearing is in place on the inside bottom of the drum throat.

LINT TRAP

The lint trap located at bottom of loading door opening MUST be in place before drying a load of clothes, *Figure 40*

Make sure the gasket around the top of lint trap is in place to prevent air leakage.

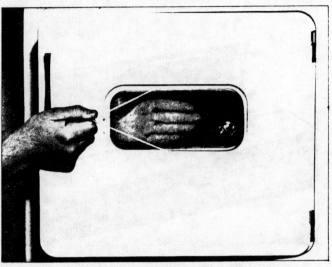

Figure 39. Gasket Replacement

NOTE: Clean lint trap prior to each load.

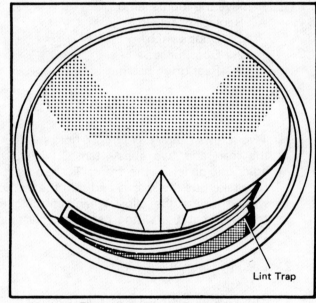

Figure 40 - Lint Collector

DOOR HINGES

Remove top panel to gain access to 3/8" nuts, securing hinges to cabinet, *Figure 38*

DOOR GLASS REPLACEMENT

To replace a broken glass panel in the dryer door, follow these steps:

1. Remove all glass fragments.

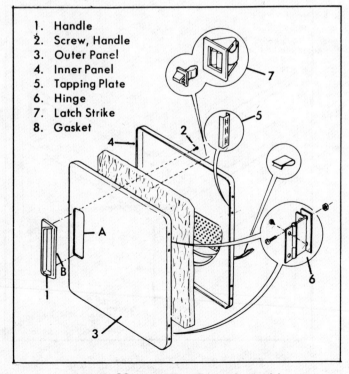

1. Handle
2. Screw, Handle
3. Outer Panel
4. Inner Panel
5. Tapping Plate
6. Hinge
7. Latch Strike
8. Gasket

Figure 38 - Loading Door Assembly

2. Remove rubber gasket from around opening. If this gasket is damaged, replace it.

3. Mount the gasket on one end of the new glass. Catch the other two corners of the glass with the gasket before slipping the remaining sides over the edge of the glass.

4. Wrap a strong cord or string once around the outer groove in the rubber gasket. Hold ends of cord firmly; do not tie.

5. Place glass and gasket firmly in position from inside the dryer door with the string ends passed through the opening to the outside. The ridges on the gasket should be toward the inside of the dryer.

6. With the gasket and glass held firmly against the door, pull ends of the string slowly to draw gasket beading over the metal edge all the way around the glass, *Figure 39.*

7. Check to see that gasket is not pinched at any point. Minor adjustment can be made by pry ing with your fingers.

LINT COLLECTOR, *Figure 40*

The lint collector rests on the front support ring assembly. The heated air is pulled from the drum and through the filter. The filter must be in place to prevent any objects, like screws or nails, from falling into the impeller.

DRUM GLIDES AND SEAL

Three front drum glides, made of Teflon, are required to support and hold the front throat of drum in position on the support ring, *Figure 42.*

The two upper drum glides are secured in place with phillips head screws. The front throat of dryer drum is supported on these bearings, no lubrication is required.

Removal of the support ring, by removing the eight screws under flange of door gasket, is required for access to the glides, *Figure 42 .*

When installing the support ring assembly, locate the bottom in first and slant it into the exhaust housing. Lift the drum up and slide the upper part of the ring into the front throat of drum and replace the screws. Be sure the spring loaded bearing at the bottom is inside the drum throat.

Figure 41 Blower Guard

The lower spring loaded bearing slides in a channel at the rear of exhaust housing, adjacent to the lint trap.

The purpose of this bearing is to hold the drum down when small loads are dried. This bearing MUST ride against the inside of drum throat to pull down on drum, *Figure 46.*

NOTE: Use high temperature wax free oil ONLY to lubricate the slide and channel.

Seal

The front drum seal, consisting of one short piece and two long pieces of felt, is fastened to the sup-

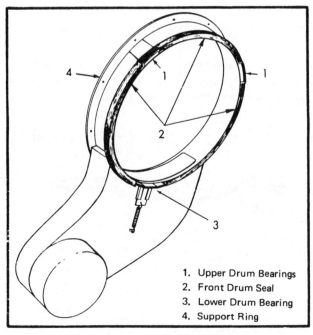

1. Upper Drum Bearings
2. Front Drum Seal
3. Lower Drum Bearing
4. Support Ring

Figure 42. - Support Ring & Housing

port ring with adhesive, *Figure 45*

The three pieces of felt seal are available separately and may be replaced as required.

The support holds the port gasket in place and keeps the front of the drum in line. To remove support ring, take out the screws and the lint filter. To install support ring, put the bottom in first and slant it into duct. Lift up the drum and slide the upper part.

UPPER DRUM GLIDES, *Figure 43*

These glides are bearing surfaces for the drum to ride on. Screws secure the glides to the support ring. To remove glides, take out the screws and work out the glides with a screwdriver. Be careful not to tear out the drum seal, *Figure 43*

LOWER DRUM GLIDES

The lower glide pulls the front of the drum down to prevent drum chatter when drying small loads. A coil spring holds the glide in place, *Figure 46*.

If the drum sticks on the glide, remove and lubricate the glide. To remove the glide:

1. Remove the front support ring.
2. Release spring holding glide in place.
3. Clean the glide.
4. Lubricate the glide with a small amount of high temperature, wax free, oil:

FRONT DRUM SEAL

This seal, *Figure 43*, is felt, 3/4 x 3/8 inches, cemented in three sections to the back edge of the support ring. The sections may be ordered separately. Order the adhesive when ordering the seal.

DRUM SUPPORT TOOL

For service procedure which required removal of the heat housing support to replace components such as heating element, drum bearings or belt replacement, drum support tool is available, *Figure 44*.

Proper support of the drum eliminates the need to reposition the front drum throat on the bearings.

Figure 43 - Upper Drum Glides

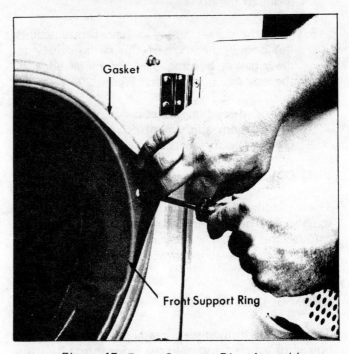

Figure 45- Front Support Ring Assembly

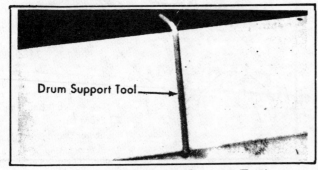

Figure 44 - Drum Support Tool

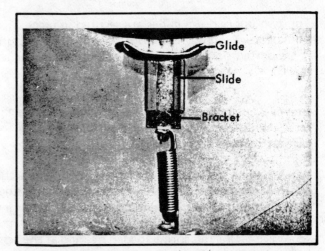

Figure 46 - Lower Drum Glide

TIMERS

EARLY MODELS

All timers used on the dryers, gas and electric, are similar, *Figure 47*. The main difference is the number of cycles and the maximum time that the timer allows for the drying cycle. They are make-and-break type switches (on-off) cam controlled and driven by a synchronous type.

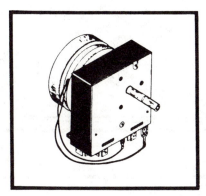

Figure 47

Electric Dryer Timers

A single cycle timer has 130 minutes for normal loads which includes a five minute no-heat or cooling period.

Gas Dryer Timers

Most of the gas dryers operates with an on-off switch type timer which has a maximum time per cycle of 130 minutes. The last five minutes is a no-heat period. During this period on the automatic ignition models, the pilot and the main burner are shut off. This makes it possible for the remaining heat to be utilized and the clothes to cool for handling. More of the heat controling timers will be explained in the gas component section.

With the exception of the SENTRY and SENSITRON controls, the timers have three contact points marked "A", "B", "C". "B" is the HOT terminal from the power line and the other two contacts provide power for the motor and the lights. Contact between "B" and "A" is the circuit for the element. Timers used with the Sentry and Sensitron circuits have an additional relay contact which is explained in the section on Dryness Control.

MOTORS

A 1/6 hp, 115 volt, 60 cycle ac motor drives the dryer at an operating speed of 1725 rpm. When facing the terminal board end of the motor the rotation is clockwise. Terminals 1 and 2 are for the motor throwout switch and the winding terminals for the motor are numbers 4 and 5. On the electric dryer the motor throwout switch is in the heater element circuit. On the gas models this switch is in the main gas valve circuit. In both cases, the switch is operated by the centrifugal force of the revolving rotor. When the motor attains the speed of approximaely 1250 rpm the motor throwout switch will close the circuit between terminals 1 and 2. If the motor speed is below 1250 rpm the motor throwout switch will be open and there will be no circuit between terminals 1 and 2.

All models are protected by a motor protector built into the motor and serviced exactly like a motor failure. If is always best to have an experienced motor repairman service the unit. It would be doing only half a job to replace the motor protector without knowing why the motor overheated. If the motor becomes overheated through being overloaded, the motor protector will cool soon after cutting out and then the motor can be restarted. Only the overload needs to be removed. If the heating is the result of bad windings inthe motor the protector will eventually fail, due to constant heat application.

MANUAL RESET DRUM PROTECTOMATIC

A safety device that breaks the circuit to the motor is the Drum Protectomactic, *Figure 48*. It opens the circuit when the temperature becomes too high for one of two reasons. Either the thermostat is incorrectly calibrated or the air ciculation path through the dryer is obstructed. The cause of the overheating should be removed before resetting the Protectomatic.

The Protectomatic device is attached at the right side of the rear drum case head. There is a small access hole in the back panel in some models and in others the top must be removed. To reset the control, depress the reset button to the locked position AFTER removing the trouble causing the "trip-out" in the first place.

Figure 48

THERMOSTATS

EARLY MODELS

The Thermostat is a-snap acting control used to control heat in the dryer. It is composed of a capillary and a bellows arrangement. The capillary is filled with a fluid, located in the drying chamber and attached to the thermostat body in the backguard. When the heat in the dryer causes the liquid to expand and flex the bellows, the thermostat contact points will open and break the circuit to the heating element. When the temperature lowers, the bellows contraction will allow the points to restablish contract and the heating element is again energized.

NOTE: The Warranty on the thermostat will be voided if the unit is disassembled. Most thermostat manufacturers will repair the unit, if returned to them out of warranty. Such information is best found in the operating manual furnished with the dryer.

Adjustment of the thermostat by operator of the dryer can be done by one of three methods. The most common is the dial indicator knob that can be turned to the proper setting. Less used but available in some models is the push button type "Fabri-Set", shown in *Figure 49*. Another style *Figure 50*, has a linkage arrangement that converts the sliding action of control knob lever to the rotary motion of the thermostat.

Sometimes the helix and control are shaft point of contact becomes rough or bent and slides with difficulty. This should be checked for and the reason removed, either by lubricating the area or adjusting the alignment of the two parts so that free and easy

action is possible. Occasionally, on older models, there will be excessive wear and the unit may have to be replaced. Other times, loosening the mounting screws and letting the linkage assembly readjust its own position will be all that is required. Be sure that the mounting screws are retightened again.

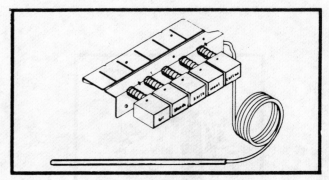

Figure 49

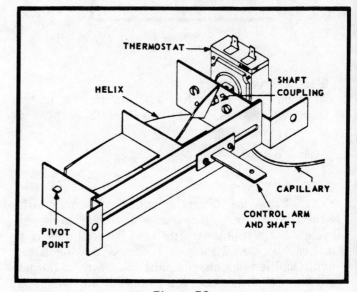

Figure 50

TEMPERATURES

Checking temperatures of the thermostat in the field is done in the following manner:

The dryer drum must have some dry clothes in it or the check will not be accurate. About 3 to 5 lbs. will be sufficient. After pulling out the lint screen, insert a long thermometer (8") into the thermometer sump shown in *Figure 51*. If there is no sump (some earlier models) it may be necessary to wire or tape the thermometer into the screen before replacing the lint screen.

With the thermostat set to its highest position, set the time for 30 minutes to allow the thermostat to cycle several times. After this time has elapsed (about three or four cycles) quickly pull out the lint screen and check the temperature which normally should be 180 degrees to 190 degrees F.

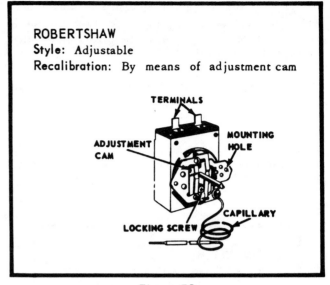

ROBERTSHAW
Style: Adjustable
Recalibration: By means of adjustment cam

Figure 52

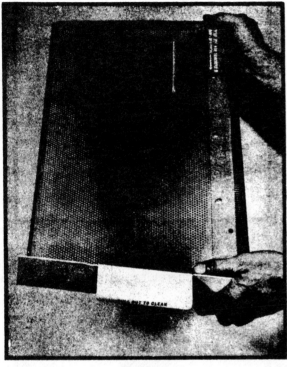

Figure 51

The use of the thermocouple is preferred over a thermometer because it is more accurate. If it is used the wires should attached to the thermometer sump near the front of the machine.

Figure 52. Expose the front of the thermostat and loosen the adjustment locking screw. Turn the notched adjustment cam clockwise to *decrease* the temperature or counterclockwise to *increase* the temperature. Only a partial turn at a time is recommended.

Tighten the locking screw and check the temperature. If it is still incorrect, repeat the recalibration procedure.

Figure 53. Remove the thermostat control knob and turn the adjusting screw, located at the base of the shaft, with a screwdriver inserted in the hollow of the shaft.

RECALIBRATING THE THERMOSTAT
All thermostats are calbrated at the factory but at times it may be necessary to recalibrate in the field. There are six models from three manufacturers and while the units are similar in action there is a slight difference in recalibrating procedures.

ROBERT SHAW THERMOSTATS
The thermostats shown in *Figure 52 and 53* are adjustable type thermostats which the one shown in *Figure 54* is a single setting type. Determine the proper model thermostat from the illustrations and then proceed to recalibrate as follows:

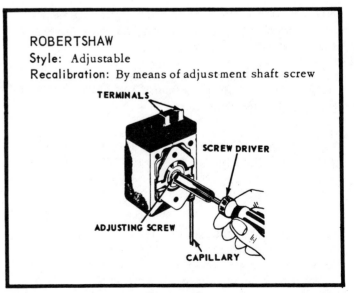

ROBERTSHAW
Style: Adjustable
Recalibration: By means of adjustment shaft screw

Figure 53

To *decrease* the temperature, turn the screw clockwise. To *increase* the temperature, turn the screw counterclockwise. Only a faction of a turn at a time is recommended.

Recheck the temperature and if not correct, repeat the recalibration procedure.

Figure 54. Expose the front of the thermostat and loosen the locking set screw. Turn the adjustment shaft screw counterclockwise to *decrease* the temperature and clockwise to *increase* the temperature.

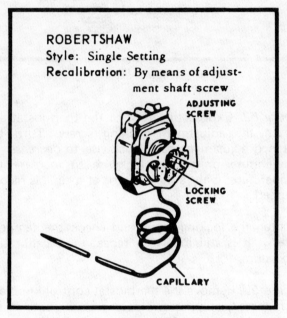

ROBERTSHAW
Style: Single Setting
Recalibration: By means of adjustment shaft screw

ADJUSTING SCREW

LOCKING SCREW

CAPILLARY

Figure 54

NOTE: *If the thermostat has not been previously recalibrated, the black seal covering the adjustment shaft screw must be cleaned out.*

Recheck the temperature and if not correct, repeat the recalibration procedure.

Two models of Wilcolator thermostats are used. Both are adjustable style thermostats. To recalibrate, find the proper model from the illustrations and proceed as follows: (Refer to *Figure 55 & 56)*

Figure 55. To adjust for a higher temperature, expose the front of the thermostat and turn the thermostat shaft to the highest heat position. Loosen the locking screw 1½ turns and then push in slightly on the thermostat shaft while turning

the shaft counterclockwise. Only a fraction of a turn at a time is recommended.

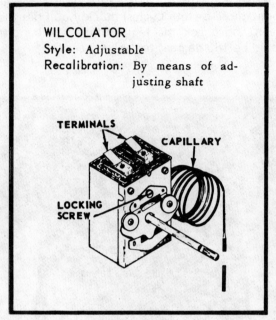

WILCOLATOR
Style: Adjustable
Recalibration: By means of adjusting shaft

TERMINALS

CAPILLARY

LOCKING SCREW

Figure 55

Retighten the locking screw and recheck the temperature. If the temperature is still incorrect, repeat the above procedure.

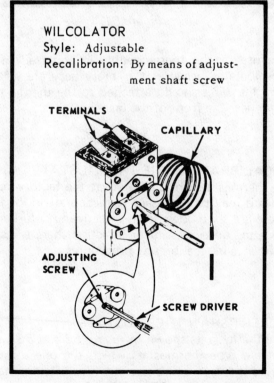

WILCOLATOR
Style: Adjustable
Recalibration: By means of adjustment shaft screw

TERMINALS

CAPILLARY

ADJUSTING SCREW

SCREW DRIVER

Figure 56

To adjust for a lower temperature, expose the front of the thermostat. Turn the thermostat shaft to the highest heat position and then turn back ¼ turn. Loosen the locking screw 1½ turns and then push in slightly on the thermostat shaft while turning the shaft clockwise. Only a fraction of a turn at time is recommended.

Retighten the locking screw and recheck the temperature. If still incorrect, repeat the above procedure.

This is an adjustable style thermostat which can be recalibrated as follows:

Figure 57. Remove the cover plate located at the rear of the thermostat. (The cover can be pried off with the fingers or a screwdriver).

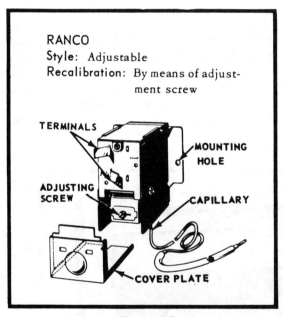

RANCO
Style: Adjustable
Recalibration: By means of adjustment screw

TERMINALS
MOUNTING HOLE
ADJUSTING SCREW
CAPILLARY
COVER PLATE

Figure 57

Now, turn the adjustment screw to recalibrate. Only a partial turn at a time is recommended. Turning the screw clockwise lowers the temperature and counterclockwise increases it.

Recheck the temperature and if correct adjustment was not obtained, repeat the above procedure.

SENTRY DRYNESS CONTROL

Two basis drying cycles are furnished with the Sentry Dryness Control called "Time Cycle" and "Auto Cycle".

"Time Cycle" can be used for heavy bulky items such as shag rugs and mattress pads, as well as small bundles (less than three pounds). The operation is this position is similar to the dryers which do not have Dryness Control. The temperature control is set for the type of load and the timer is set for the desired drying time. When the selected time has elapsed, the dryer will shut off. There is a no-heat or cool-down period of 5 minutes included in the drying time.

The "Auto Cycle" is for all average loads. The dryer will shut off when the load has reached the degree of dryness that has been selected. The temperature control is set at the desired temperature setting. "A", "B", and "C" settings are available for the type of load and the degree of dampness desired.

These settings are explained in the Operators Instructions. All of the drying is accomplished through this relationship of the timer and the thermostat.

The timer will not advance until the thermostat has cycled and does not call for heat. This is accomplished by the use of a relay in the circuit. When the thermostat has cycled, the relay circuit is opened. As a result, the switch of the relay closes thereby completing the timer circuit. See *Figures 58 and 59*. When the thermostat again calls for heat the timer motor circuit is opened. As this "on" and "off" cycling continues for the predetermined time, it results in the degree of dryness selected. After the heating cycle is completed, there is approximately 6 minutes of no-heat or cooling off period and the dryer shuts off.

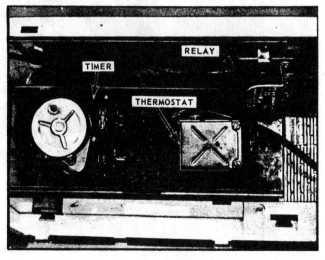

RELAY
TIMER
THERMOSTAT

Figure 58

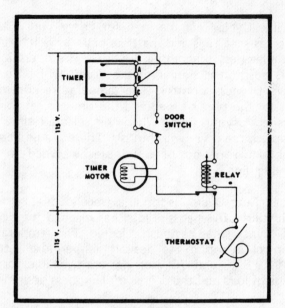

Figure 59

The AUTOMATIC Dry Cycle uses a filtered dc circuit which originates in the dryness control and is directed to the sensing rods in the dryer drum. The sensing rods are so located as to be in constant contact with the drying load where the wet clothes act as a conductor which completes the cirucit to the ground.

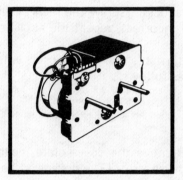

Figure 60

THE SENSITRON DRYNESS CONTROL

(Used on BOTH electric and Gas Dryers.)

The Sensistron Dryness Control, shown in *Figures 60 and 61,* is basically an electronic device. However, it does incorporate a TIME DRY cycle which the operator may set in the conventional manner. The timer and the thermostat control the time dry cycle in exactly the same manner as they did in the Sentry Control explained previously.

When the moisture content of the load is reduced to the desired degree of dryness (as set on the dryness control) the circuit is broken. Then, in accordance with the setting chose, the dyrer control will either shut off the dryer or start a predetermined time cycle. See *Figure 62* for the details as to how settings are made.

Figure 61

SENSITRON CONTROL SETTINGS AND FUNCTIONS CHART			
Cycle	Temperature Control (Left Knob)	Automatic and Time Dry Cycle Control (Center Knob)	Dry – Dampness Control (Right Knob)
T I M E D R Y	**SETTING:** Operator turns knob until indicator points to "Air", "Lo", "Med", or "Hi" setting. **FUNCTION:** Controls temperature of drying chamber. When degree of heat is reached, control interrupts heat circuit. Circuit remains open until temperature lowers.	**SETTING:** Operator turns knob (in TIME DRY section of dial) until indicator points to the selected number of minutes of desired drying (up to 60 minutes – last 5 minutes which are a no-heat period.) **FUNCTION:** Controls length of drying cycle by means of switch contacts operated by a timer motor driven cam.	This control setting is disregarded as moisture sensor is not utilized in TIME DRY cycle.
A U T O M A T I C D R Y	**SETTING:** Operator turns knob until indicator points to "Air", "Lo", "Med", or "Hi" setting. **FUNCTION:** Controls temperature of drying chamber. When degree of heat is reached, control interrupts heat circuit. Circuit remains open until temperature lowers.	**SETTING: (For loads over 3 lbs)** *IMPORTANT – SET SENSITRON DRY-DAMPNESS CONTROL (RIGHT KNOB) BEFORE SETTING THIS CONTROL.* Operator turns knob (in AUTOMATIC DRY section of dial) until indicator points to either the "Wool", "Normal" or "Heavy" setting (see function information below for differences.) **NOTE:** Closing the door in the "Wool" cycle does not re-start the dryer as it does in all other settings. To re-start the dryer, close the door and then rotate the knob indicator through the "Wool" dial mark to re-energize the electrical circuit. **FUNCTION:** Controls length of operation by actually measuring the moisture content of the clothing being dried. When the selected degree of dryness is attained, the dryer will shut off. The difference in settings is as follows: ••• "Wool" – Sensing controlled heat period – no cool down period. ••• "Normal" – Sensing controlled heat period – plus 7 minutes cool down period. ••• "Heavy" – Sensing controlled heat period – plus a 13-minute timed heat and a 7-minute cool down period. (Designed specifically for extra heavy garments, such as shag rugs, mattress pads, etc.)	**SETTING:** Operator turns knob until indicator points to either "Dry" - for shelf storage (3 to 5% moisture) or from "Damp" to "Max-Damp" setting - for ironing (15 to 40% moisture.) **FUNCTION:** Controls degree of dryness. When degree of dryness is reached, the sensing circuit signals the dryness control.

Figure 62

The Sensitron check-out chart in the front of this manual should be followed very closely in order that a true check can be determined.

FUSE LINK REPLACEMENT

The early model dryers have a fuse link that is used as an extra safety device. When the temperature reaches an excessive limit, the fuse will melt and must be replaced before the circuit will close again. The circuit is to the solenoid valve which turns off the gas to the main burner.

To replace this fuse, the hand-hole cover at the back of the dryer is removed and the fuse wiring disconnected. By removing the screws that hold the fuse link body to the cabinet, the fuse can be replaced. The Long terminal is placed UP so that the loop is suspended. This fuse is designed to "blow" only after the thermostat exceeds 190 degrees F. A good chance exists, when it blows, that something else is wrong and the reason should be check out. *Figure 63.*

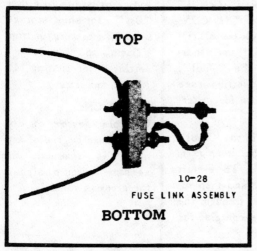

Figure 63

HEATING ELEMENT REPLACEMENT

The heater coils, of chrome-nickel alloy called nichrome, are suspended in heat resistant ceramic bushings, the ends are brazed to the terminal studs to insure positive connection. The studs are inserted into a special locking type ceramic block and may be removed in the following manner:

Turn off all power and remove the back panel. Removing the terminal leads as shown in *Figure 64.* The screw holding the heater clip, *Figure 65,* can be taken off and the heater slid out carefully, as shown in *Figure 66.* Take out two screws located at the end of the heater and remove the screen protector after loosening the screen tabs.

To replace the unit, reverse the above procedures. Be sure that the same wattage element is replaced, 4400 watts for 230 Volt operation and 1100 watts for 115 volt opration.

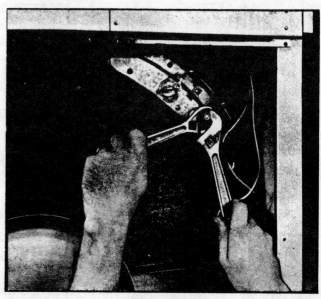

Figure 64

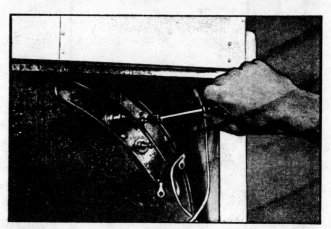

Figure 65

Figure 66

SECTION 3

SERVICE PROCEDURE MECHANICAL COMPONENTS

Diagnosing mechanical problems is generally done by a visual inspection. Understanding the functional operation of the various mechanical components is necessary, in many cases, if the proper diagnosis is to be made. Knowledge of what these components do and then observing them in their position and/or removed will readily reveal any malfunction.

Study the following section carefully, associating the various components with the illustrations and text. Much time can be saved repairing machines if this functional data is clearly understood.

Due to the large number of models covered in this manual, no attempt will be made in this section to give a complete detailed step-by-step procedure on disassembly of each individual model. Instead, we will give the service procedure and functional description of the various components as used on most models. In a few cases these components may not be identical to the machine being serviced but their function, as well as service procedure, will be the same.

TWO DRIVING MECHANISM, EARLY MODELS

To drive the drum within the drum case, as well as the blower assembly, a system of belts and pulleys is used. To provide tension for the belts, an idler pulley is mounted on the idler bar and kept in the correct position by a spring as shown in *Figure 67.* The idler bar fits into a slot, detailed in *Figure 68,* the late model version. Prior to serial number 00205-420, the retaining ring was not used.

The change was made to control the position of the idler pulley on the idler pulley bearing. Both types are mounted on the upright channel of the rear drum case band.

MOTOR-MOUNTED BLOWER ASSEMBLY

The air movement as shown in *Figure 69,* is provided for by the motor and blower assembly which can be removed as assembly after the belt has been removed and the motor saddle securing screws removed as shown in *Figure 70.* Don't forget to disconnect the motor leads. The motor and blower can be then removed by sliding it straight back out of the dryer, see *Figure 71.* (The whole mechanism can be serviced after the cabinet back and the exhaust extension is removed.) Most of the mechanical work is done from the rear of the machine.

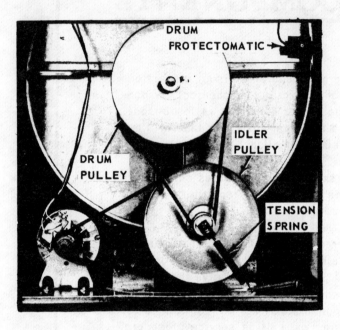

Figure 67

Figure 70

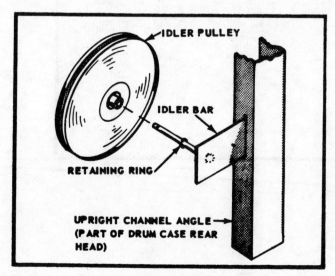

Figure 68

Figure 71

Air Flow — Gas Dryer

OPERATION

The dryer will operate when it is connected to the proper electrical power and gas source. The door switch must be depressed. The timer knob must be turned on and in models with a momentary contact switch the knob must be pressed in to start. The power goes from the timer to operate both, the motor which turns the drum and the blower fan, and the heat source. The drum and he impeller will turn as long as there is time left to operate as indicated on the timer. The heat source is controlled by a limiting thermostat which can be adjusted in most models by a heat selector switch. A safety thermostat protects the heating system if the limit-in thermostat is defective.

Air Flow — Electric Dryer   *Figure 69*

Pry off Four Metal
Clips Holding Two Sections of
Fan Housing Together.

Figure 72

Loosen Set Screw
Holding Fan to Motor Shaft
and Remove Fan.

Figure 73

Remove Four Screws
Holding Fan Housing to Saddle
and Remove Housing.

Figure 74

Four screws hold the blower assembly to the motor saddle. By prying off the metal clips around the two sections of the fan housing, the housing will separate so that the set screw holding fan to motor shaft can be removed. The four screws holding the blower housing can now be removed and the motor will be free. *Figure 72, 73 and 74* illustrate the disassembly procedure. To reassemble, the above steps are reverse.

When reinstalling the fan, it should be positioned on the motor shaft so that the minimum clearance is obtained on the intake side of the fan housing. When tightening the cup point set screw, exert enough pressure to allow the cup point to bite into the shaft. A heavy-duty allen wrench will be needed, with a possible assist from pliers carefully applied. If the set screw is of the double locking variety, (part numbers 14-162 and 14-163) the locking set screw must be drawn up tightly to the cup point set screw.

Because the housings are of precision fit, care shown shown in assembly will insure the maximum efficiency available from the repaired unit. The six-vaned fan is designed to exhaust air at the rate of 90 cubic feet per minute. It must be installed without bind or bad alignment. Particular attention must be paid to the positioning of the fan. Some of the earlier models were metal and some of Phenolic plastic.

After serial number 00208-402, phenolic fan with sintered iron hub was used to equip all dryers. This fan can be used to replace all prior model fans. It has a single set screw, part no. 14-150.

When a die cast fan is used, with a single set screw, better tightening on the shaft can be accomplished by the use of a Neoprene plug, part number 11-88. It should be inserted into the set screw opening before the set screw is turned in. The pressure of the set screw will be improved when the plug is forced between the fan hub and the motor shaft. See *Figure 75. Figure 93* shows the correct positioning of the fan hub for the particular models of gas dryers that are identified on the drawing. These tolerances should be observed in order that the fan not bind or be distorted by the heat.

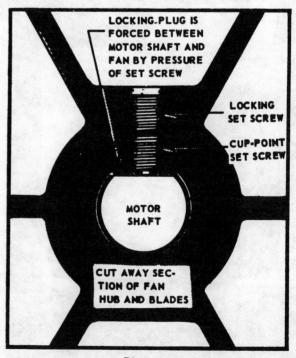

LOCKING PLUG IS
FORCED BETWEEN
MOTOR SHAFT AND
FAN BY PRESSURE
OF SET SCREW

LOCKING
SET SCREW

CUP-POINT
SET SCREW

MOTOR
SHAFT

CUT AWAY SEC-
TION OF FAN
HUB AND BLADES

Figure 75

BELT

The single poly V belt is routed around the drum and motor pulley, then over the idler pulley. Tension is

maintained on the belt by a spring which attaches to to a support gusset at left side (facing back of cabinet) and to the idler pulley arm, *Figure 77*

The spring loaded idler wheel and arm assembly is held in place on dryer base, idler arm in nylon bearing, with a bracket fastened to base with a sheet metal screw.

The flange on wheel is towards the front of dryer.

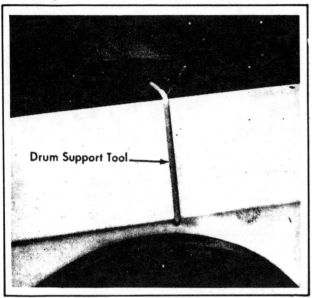

Figure 76 - *Drum Support Tool*

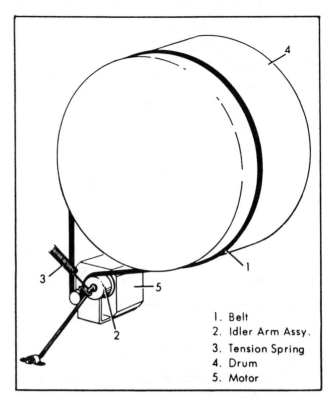

1. Belt
2. Idler Arm Assy.
3. Tension Spring
4. Drum
5. Motor

Figure 77 Belt Routing

Belt Installation

1. Install the drum support tool.

2. Remove rear cabinet panel.

3. Remove light and harness connections, as required.

4. Remove bolts from heat housing support, each side of cabinet, and two screws which fasten heat housing to strut at top.

   NOTE: Screw which fastens heat housing to duct must be removed on gas dryers.

5. Locate belt on drum and pulleys and reassemble all parts except rear panel, then remove support tool.

6. Check front support ring to be sure the bearings are in place in drum throat.

7. Align belt on drum and pulleys, *Figure 77*

8. Operate dryer and observe belt. Be sure it rides in position before installing back panel.

   NOTE: Belt may be installed from the front of dryer by removing the support ring. This method also requires removal of the rear panel for installation and alignment of belt on pulleys and drum.

DRUM REAR BEARING, *Figure 78*

The self-centering bearing, located at the center of heat housing support, is replaced as follows:

1. Support drum to hold it in front support ring.

2. Remove snap ring from drum shaft. Do not over stretch snap ring, *Figure 78.*

   CAUTION: Be sure to note the location of all fiber and metal spacers for reassembly.

3. Remove bolts from heat housing support which hold bearing wick and collars in place.

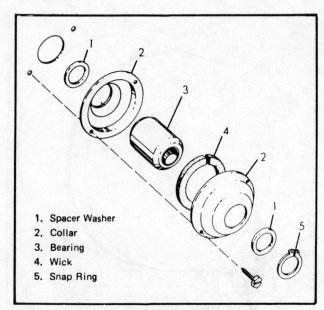

1. Spacer Washer
2. Collar
3. Bearing
4. Wick
5. Snap Ring

Figure 78. - Drum Rear Bearing

DRUM, Later Models

The drum is supported at front by the support ring assembly, and at the rear by a removable shaft held in the drum bearing by a snap ring.

Drum Removal

1. Remove top panel, loosen harness from clip, and carefully hang top on side of dryer cabinet. Use care not to scratch or mar the cabinet.

2. Remove front support ring.

3. Use a padded surface on floor to protect cabinet front and door handle. Lay dryer on its front, carefully supporting top and set top on floor next to dryer.

   NOTE: Two 2 x 4 support blocks, 4-1/2" long, may be located between front edge of drum and inside center of cabinet front, to prevent drum from dropping when snap ring is removed.

4. Remove cabinet rear panel.

5. Remove belt from pulleys.

6. Remove screws which secure heat housing to combustion chamber, on gas dryers only.

7. Remove strut at top rear of cabinet.

8. Remove light and harness connections to components, as required.

9. Remove bolts from heat housing support at each side of cabinet, and remove snap ring.

10. Lift heat housing from cabinet, *Figure 79*

11. Drum shaft, drum, or rear drum seal are now accessible for replacement.

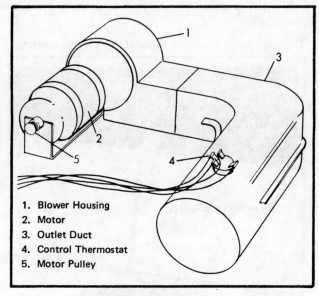

1. Blower Housing
2. Motor
3. Outlet Duct
4. Control Thermostat
5. Motor Pulley

Figure 80

DRUM SHAFT

The tempered shaft, with right hand threads, fastens into the rear plate of dryer drum. A lock washer prevents the shaft from loosening.

Refer to drum removal for shaft replacement procedure.

BLOWER IMPELLER

The blower impeller, equipped with an insert which has left hand threads, is mounted on the motor shaft, See *Figure 80*.   *Figure 81*.

HEAT HOUSING

For the removal of the heating element from an electric dryer, the heat housing must be removed.

To remove the housing:

1. Secure the front of the drum to the front glides as described *in Figure 79.*

2. Remove the back access panel.

3. Remove the snap ring from the drum shaft.

4. Remove the four bolts and nuts securing the housing to the cabinet and the two screws holding the housing to the top cross truss.

Figure 79. - *Heat Housing*

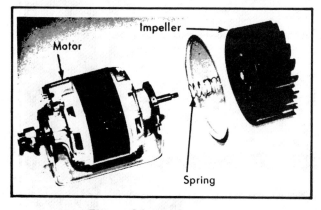

Figure 81 - Blower Impeller

Rear Access Panel, Later Models

Removal of the back panel gives access to the drum assembly, motor and blower fan, electric heating element, drum lights and safety thermostat.

IDLER ARM ASSEMBLY

The idler arm assembly, *Figure 83,* which includes the arm, spring and wheel, keeps the drive belt tight around the drum and motor pulley. A mounting bolt holds the arm to the dryer base and the spring, anchored to the cabinet brace, pulls the arm to keep

the belt tight. To release the drive belt, remove the spring from the arm with a pliers, *Figure 84.* The idler wheel can be replaced. Remove the snap ring with a tru-arc pliers. The flange on the wheel is toward the front of the cabinet, *Figure 82.*

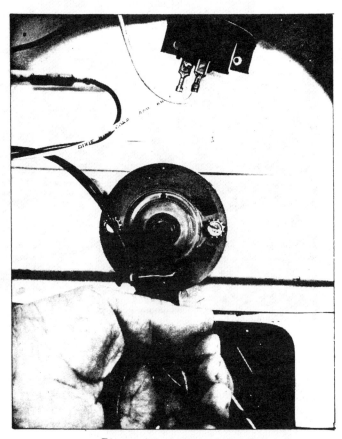

Figure 82 - Heat Housing

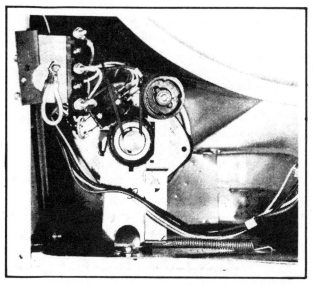

Figure 83. - Idler Assembly

Figure 84. - Idler Arm Assembly

DRIVE BELT

To inspect a drive belt, lift up the top panel.

To replace a burned or broken belt:
Gas Dryer Only

1. Remove the screw holding the combustion chamber to the heat housing.

2. Insert the new belt under the heat housing.

3. Flip up the combustion chamber

For Both Gas and Electric

4. Remove electrical connections from the heat housing. *Figure 85.*

5. Remove the mounting bolts and nuts holding the heat housing on the right side and insert the belt, *Figure 86.*

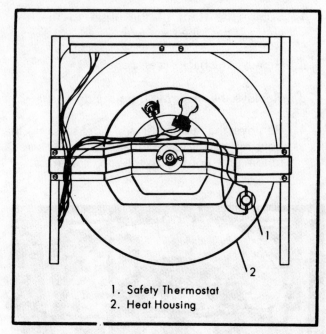

1. Safety Thermostat
2. Heat Housing

. *Figure 85. Location Safety Thermostat Gas Dryer*

Figure 86 - Heat Housing Mounting Brace

THE DRUM SHAFT

The shaft, *Figure 87.* connects the drum to the cabinet assembly. The shaft can be removed. To remove the shaft, follow instructions for removal of the drum. The shaft has a normal right-handed thread.

Figure 87 - Drum Shaft

REAR DRUM SEAL (GAS DRYER), *Figure 88*

The rear drum seal prevents air leakage around the back of the drum. If the seal becomes defective, order kit F61935. This kit includes:

1	A2580	Fiber Seal
1	A518	Spring
1	55624-2	Adhesive, (1 oz.)

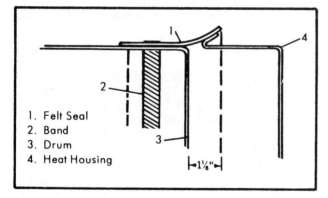

1. Felt Seal
2. Band
3. Drum
4. Heat Housing

|←1⅛"→|

Figure 88 - Gas Dryer Rear Drum Seal

To install a new seal:

1. Follow instructions for removal of the drum.

2. Locate the metal clip holding band to drum.

3. Drill two 1/8 inch holes, 1 - 5/8 inches apart on either side of the clip; do not drill through drum.

4. Cut clip from the band.

5. Remove the old seal.

6. Clean one inch section around drum. Remove any oil or foreign matter.

7. The seal is two inches wide. To insure that the

seal covers the edge of the heat housing, the seal should project over the edge of the drum by 7/8 inch.

8. Apply adhesive to the last one inch of the drum.

9. Slide the seal over the drum.

10. Position the steel band over the seal.

11. Install spring into holes.

12. Follow instruction for installation of the drum.

EXHAUST SYSTEM

The exhaust system includes, *Figure 89.*

1. The blower and motor.

2. The blower housing.

3. The outlet duct.

Figure 89 - Air Duct and Blower Exhaust

The impeller fan, *Figure 90,* driven by the motor, pulls the air from the drum and into the outlet duct. The blower housing has a rubber seal in the front and a cover plate behind to prevent air leakage.

The motor, cover plate and impeller come out as a single unit. To remove the impeller fan, the motor must be removed. A screw secures the motor to the cabinet in the rear and a tang holds the mount at the front, *Figure 90.*

The spring loaded blower housing cover closes the opening in housing when the motor is installed in place.

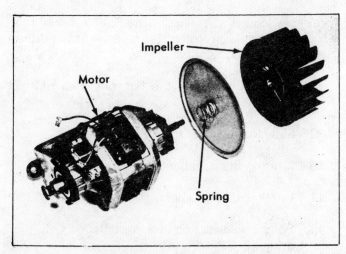

Figure 90 - Motor and Impeller

Remove the rear panel and motor for access to the blower impeller.

OUTLET DUCT

The outlet duct connects direct to the blower housing and the exhaust pipe.

The duct is fastened to the dryer base with screws installed from the inside, down through the base.

The control thermostat is mounted on the outlet duct, *Figure 91*.

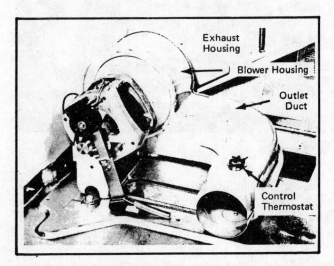

Figure 91 - Outlet Duct

REAR DRUM SEAL, REPLACEMENT

Gas Dryers Only

This seal is required on gas dryers so all heated air from the combustion chamber will be drawn through the heat housing into the drum.

The seal must be in good condition to prevent room air from mixing with the heated air, otherwise, slow drying will be the result.

The seal kit consists of a seal spring and one ounce container of adhesive.

Seal Installation

1. Turn drum so clip on steel band is up.

2. Mark and drill two 1/8" holes in steel band, 1-5/8" apart, one hole at each side of clip. DO NOT PERMIT DRILL TO ENTER DRUM.

3. Cut steel band as close to each side of clip as is possible.

4. Remove the old seal and clean the seal bonding area thoroughly. Remove as little of the paint as possible.

5. Apply adhesive around drum at area of seal contact. See *Figure 92*.

6. Slide the new seal on the drum and position it over the area coated with adhesive.

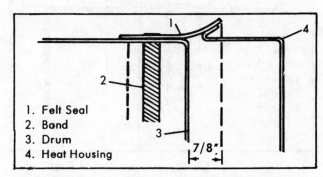

1. Felt Seal
2. Band
3. Drum
4. Heat Housing

Figure 92 - Rear Drum Seal - Gas Only

To remove the motor and impeller fan:

1. Release the spring holding the idler arm.

2. Disconnect the harness at the motor.

3. Remove the screw holding the motor mount.

4. Pull up and out on the motor.

5. Rotate the assembly to remove it from the dryer.

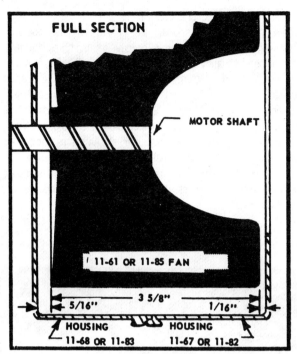

Figure 93

SENSITRON MODELS

Contact and slip ring assemblies are used on all SENSITRON dryness controlled dryers. These assemblies provide a way for the circuitry of the dryness control to be positively transfered from the Sensitron unit to the sensing rods, by way of the drum spider shaft, so that it is completed back to th ground of the machine.

The contact assembly is fasten to the horizontal cross angle of the rear drum case head by two screws. See *Figure 94*. There is a carbon button designed to make contact with the revolving slip ring. This contact must be constant and sure at all times. Therefore the tension, which is slight, must be adjusted at times by loosening the screws on the rear cross angle and moving the assembly into position of better contact. Retighten the screws and check the carbon button contact by turning the pulley and observing that the contact is constant for the full revolution of the slip ring.

Any noise or squeaking of the carbon button can be removed by the application of a thin coating of an approved conductive lubricant.

To replace the contact ring, disconnect the lead from the slip ring tab and the wire from wire clamp on the drum pulley. Then the pulley and belt can be removed by loosening the set screw on drum pul-

ley. Slide the slip ring off the spider shaft. Replace in reverse order.

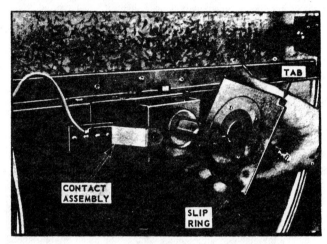

Figure 94

SENSING PLATES REPLACEMENT

In all dryers using the Sensitron dryness control, there are three sensing plates attached to the drum baffles, *Figure 95*. They are held in position by two screws, passing through ceramic insulators to the drum baffle. They are insulated from the baffle by a plate insulator and connected to a wire which passes through the rear drum head; led to the center of the drum head where they are soldered together with the lead wire from the slip ring assembly. To avoid accidental contact with the rear drum head case, the wire leads are taped to the spider arms with fiber tape, *Figure 96*. Certain models may have special speed clips to hold lead segments.

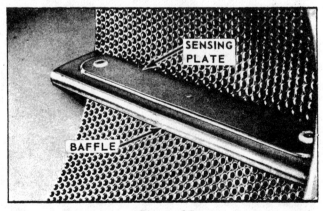

Figure 95

To replace a sensing plate that may be defective, turn off all power and remove the screws and ceramic bushings that hold the sensing plate to the baffle. By turning the sensor plate over, the wire lead can be exposed. This should be cut close to the terminal connection on the plate and the removed lead stripped back about 1/4 inch. Cut off the replacement plate lead until approximately 4 inches remain on the plate. After stripping this lead to 1/4 inch from end, place the fiberglass insulator over the wire and splice the wires together and solder, *Figure 97.* Wrap the connection with fiberglass tape. The excess wire can be pushed back through the hole in the baffle and the insulator and plate reassembled.

Figure 96

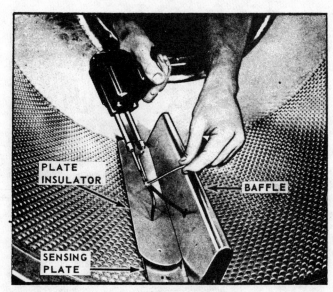

Figure 97

DRUM CLEAN-OUT COVER.

Clean out covers are in all dryer drums to aid in cleaning out the accumulation of lint, because the lint screens may have neglected. To remove the lint, turn the slotted locking device and remove the cover. Then the lint can be brushed into the lint screen area and removed. It will pay to do this every service call to eliminate excessive pile-up of lint on the wrapper sheets. *Figure 98* shows the location in the drum. Some models have another clean out cover on the rear of the drum that will enable a special brush (furnished with the dryer) to be used. *See Figure 99.*

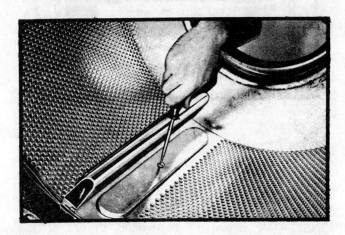

Figure 98

Figure 99

TO REMOVE THE DRUM.

Remove the back panel and top assembly. As demonstrated in *Figure 100,* the rear cross angle assembly can be moved to a position above the dryer (where it can rest on the drum wrapper) after the screws are removed that fasten it to the cross bar.

Figure 100

The wires to the motor, heater drum protectomatic, and the slip ring (if one is used) should be disconnected at the terminals. Slip the drive belts off and remove the heater, pulleys, spring and idler bar. There are two screws holding the case head upright to the base, *Figure 101,* when removed and the rear drum retainer bolt loosened, *Figure 102* the drum case head can be pryed out with the aid of two screwdrivers operated as shown in *Figure 103.*

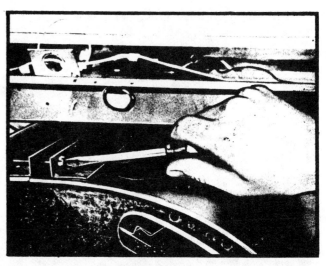

Figure 102

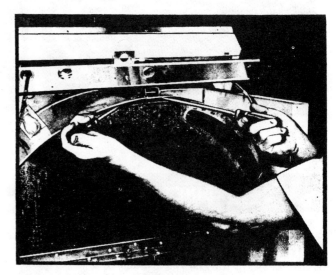

Figure 103

Figure 101

Grasping the case head by the cross arm and lifting up will slide the whole assembly from the spider shaft if care is taken to see that tabs of the case head clear the wrapper sheet slots. Shown in *Figure 104* is the correct way to grasp the assembly. After this, there is no trouble to slide drum out of the drum oven, *Figure 105*.

CASE
HEAD
TABS

Figure 104

To reassemble, reverse the order of the instructions just given.

Figure 105

CABINET SERVICE

EARLY MODELS

In some Hamilton Dryer models the controls can be serviced by the removal of the complete cabinet top, others have the controls uncovered by simply tilting the backguard forward. Some models require the top to be removed as a unit, after backguard is tilted to the proper angle to clear the top as it is removed.

Certain models such as the one pictured in *Figure 106,* have a mechanism that releases when the timer shaft and thermostat shaft are pushed downward and to the rear. This releases the whole top to be lifted off as it is moved slightly backward and upward to disengage top flange from dryer front retaining lip or flange.

Most of the recent models have a backguard that tilts forward. The only other concern is to remove either shipping screws from backguard or to release clips. A few of the dryers, using shipping screws, have a backguard that must be removed with the top after removing the knobs. The cabinet

top is then pulled upward to disengage it from the retaining clips and then pushed backward from top after which the front flange will be cleared. The top then lifts off.

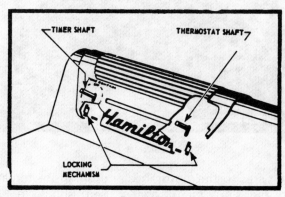

Figure 106

Certain few models may release the tope by depressing the timer shaft ONLY, in a downward direction while also moving it backward (toward the rear of cabinet).

Always use a soft cloth protector between the backguard and the cabinet top to avoid damage to finish.

LOADING DOORS

Loading doors in all models have two hinges. Lock washers are used to insure alignment and if left out may result in a small "dimple" on the outside of the door.

There is a friction catch on the door and the door pull is attached from inside the panel.

To replace the door glass, remove all broken glass and the rubber gasket from the door. If the gasket is damaged, it should be replaced. Before trying to install a new glass in the door, place one end of it in the new gasket and catch the other corners in the gasket groove before slipping the remaining side over the edge of the glass.

Wrap a strong string around the outer groove in the gasket and hold the ends firmly. Do not tie.

Place the glass and gasket firmly in the door opening, as much of it as will go, with the strings passed through the opening. The strings can be drawn slowly and firmly so that the beading of the gasket will be fed over the metal edge of the frame and around the glass. *Figure 107* is an example of the proper positioning. Make sure that gasket is not pinched or torn in the process. It can be positioned more neatly with the fingers after being fed into position.

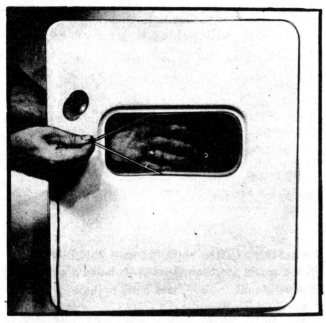

Figure 107

DOOR SEALS.

Loading door seals (port rings) are of two varieties. One is on the door and the other is on the cabinet port opening.

The loading door using a seal is a double panel door. It has the sealing ring and the gasket in the door and seals against a phenolic plastic ring. The gasket can be removed by taking out four sealing panel screws and pulling it out.

A single panel door is used on the models that have the port ring in the cabinet. The port ring is actually a combination seal and port ring. The ring is attached to the cabinet front and over the port opening to the drum case flange. *Figure 108* shows the proper installation method. A cross-section view of the ring is shown in *Figure 109*.

To replace the port ring, remove the old one by grasping front lip of the seal and rolling it toward the center of the ring opening far enough to clear the cabinet front. Now, the ring can be pushed into the drum. The new one can be inserted and the top held against the top of the opening, while the other hand grasps the bottom of the ring. Double it up to take surplus material under control as shown in *Figure 110*. Starting at the top, work the channel of

the seal over the drum case flange and then redistribute the surplus material to obtain equal pressure. Check to see that it engages the drum case flange at all points. If it is not installed carefully it can hold the door open or cause a rubbing of the revolving drum.

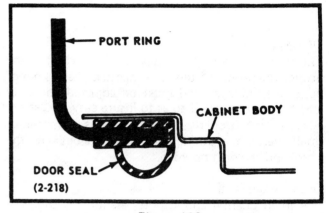

Figure 108

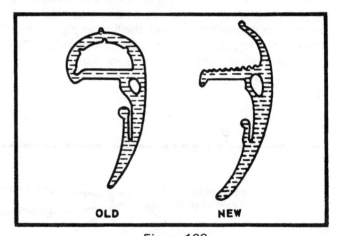

Figure 109

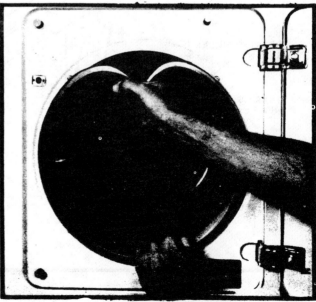

Figure 110

Depending on the model, three styles of port rings are used. Two of them are Phenolic rings with the Sun-E-Day bulb and shield mounted on one of them while the other one is without this light. The third type of ring is made of Elastimer and does not have the Sun-E-Day light, either.

DRUM DRIVE.
The drum and drum case service is important to the proper operation of the dryer because the system of belts and pulleys used must be adjusted correctly and the drum aligned so as to insure easy operation. Also, if not properly aligned there is a danger of small pieces of clothing getting caught between the drum and the port opening.

Drum Alignment.
The main bearing and U-bolt assembly is the controlling adjustment point for the alignment of the drum. The U-bolt relationship to the main bearing is shown in *Figure 111.* Both the older and the newer types are shown and the following instructions apply in each case: First, check the spacing between the loading port and the drum flange to determine alignment needs.

Older Stlye. (Prior to serial number 20917-001)
If the drum position is too high, U-bolts should be tightened or shims removed evenly until the correct position is obtained.

If the position is too low, loosen the U-bolts and place a bearing shim between the filler angle and the angle bracket. Washers should not be used because they will not exert pressure correctly.

The right side of the dryer drum (as you face back of dryer) may be too tight. If so, go to the back of the dryer and tighten the right side nut as you loosen the left side nut. Remember that you are now facing new right and left side positions — your right and left.

If the left side is too tight, facing front of dryer, reverse the tightening and loosening procedure at the back of dryer.

Newer Style. (After serial number 20917-000)
If the drum is too high, loosen U-bolts and adjusting screw lock nuts. Turn both adjusting screws IN an equal amount to secure correct position. Make turns slowly and recheck! Be sure and retighten the U-bolt and the adjusting screw lock

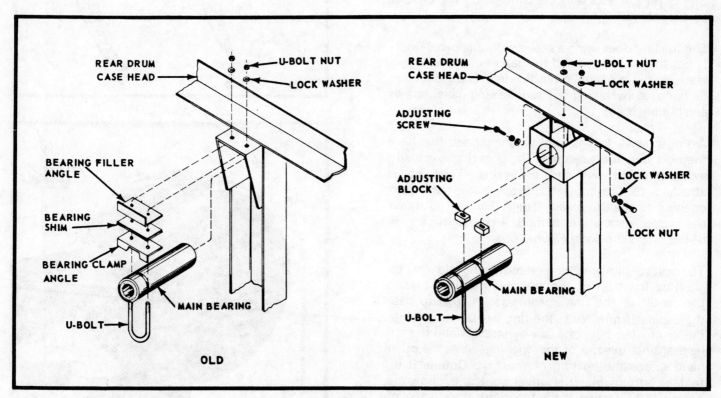

Figure 111

nuts after securing alignment. If the drum is too low, turn OUT both adjusting screws and tighten both U-bolts an equal amount. When correct alignment is obtained, tighten both adjusting screws and lock nuts.

If the right side is too tight (facing the front of dryer) loosen the U-bolt nuts and adjusting screw nuts, facing the back of the dryer turn the right adjusting screw out and the left screw in, until adjustment is correct. To adjust the tight left side, reverse the procedure: left adjusting screw out and the right adjusting screw in. Retighten all nuts.

VENTING

Because the moisture laden air being exhausted from the dryer is often objectionable, attention should be paid to this conditon. If the dryer is to exhaust into the room where it stands, the cross ventilation between windows, or a window and a doorway, may be satisfactory. However, in most cases that the serviceman will encounter, this method is not to the best advantage of the occupants of the area, or conductive to good drying of the clothes.

Venting should be done as near to an outside wall as possible in keeping with the rule to have the outlet pipe as short as possible. Exhausted dryers can be installed close to walls or other cabinets but the unexhausted dryers should always have at least 4" of clearance away from other cabinets (or the wall) at the rear. Some models exhaust from the rear and require no special deflector kit. Others will require special kits to attach the outlet pipe to for the best results. *Figures 112 and 113* show the two exhaust systems used. One is called the "Low Pressure" front exhaust and the other the "High Pressure" type, installed at the rear of the cabinet. Most of the present day models are high pressure types. Kits for the adaptation of the outlet pipes are shown in *Figure 114.* Typical installations are also shown for the various locational positions and type of room that may be the encountered. Instructions are available with the kits to make the correct connections but the following general instructions are important to observe.

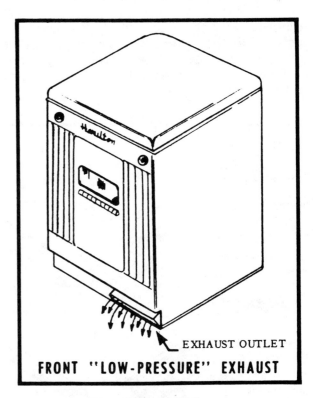

FRONT "LOW-PRESSURE" EXHAUST

Figure 112

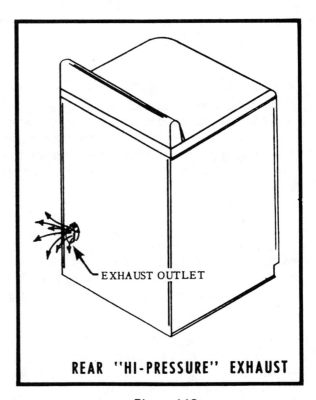

REAR "HI-PRESSURE" EXHAUST

Figure 113

After locating the area best suited to the needs of the dryer, cut the holes in the walls, windows or other places of outlet, and observe the limits of exhaust lengths of pipe and the proper elbows and fittings to carry the vapors and lint away without clogging. No sharp turns or rough edges inside of the pipe should be tolerated. They only restrict the flow of air and form pockets of lint that can not be carried out of the pipe.

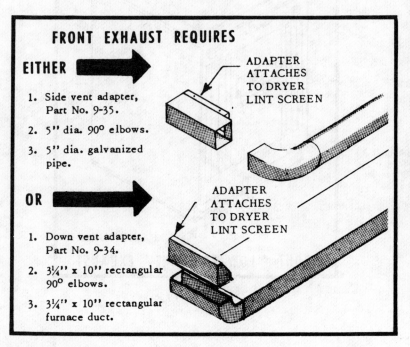

FRONT EXHAUST REQUIRES

EITHER ➡

1. Side vent adapter, Part No. 9-35.
2. 5" dia. 90° elbows.
3. 5" dia. galvanized pipe.

ADAPTER ATTACHES TO DRYER LINT SCREEN

OR ➡

1. Down vent adapter, Part No. 9-34.
2. 3¼" x 10" rectangular 90° elbows.
3. 3¼" x 10" rectangular furnace duct.

ADAPTER ATTACHES TO DRYER LINT SCREEN

Figure 114

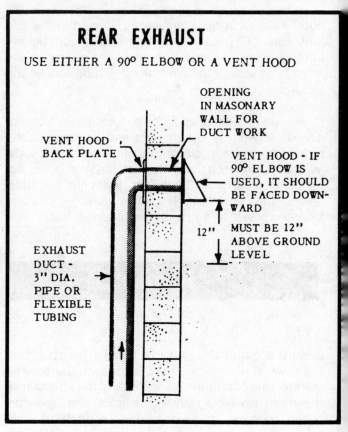

REAR EXHAUST
USE EITHER A 90° ELBOW OR A VENT HOOD

VENT HOOD BACK PLATE

OPENING IN MASONARY WALL FOR DUCT WORK

VENT HOOD - IF 90° ELBOW IS USED, IT SHOULD BE FACED DOWNWARD

EXHAUST DUCT - 3" DIA. PIPE OR FLEXIBLE TUBING

12" MUST BE 12" ABOVE GROUND LEVEL

Figure 115

Draft caps over the outlet on the outside of the building should be provided, either as shown in *Figure 115* or improvised such as the inverted elbow in *Figure 116*. Needless to add, the connections and fittings should be made tight. Pipes smaller than allowed by local code (or good usage) should not be installed.

The pipe should be of rust-inhibiting materials so as to cut down the chance of early disintegration of the exhaust outlet. Many times the line runs in such a way as to be unnoticed for long periods of time and so escape detection when rusted through.

The outside vent hood should always be at least 12" above ground. NEVER vent the dryer into any other gas vent arrangement from another appliance. A vent into a chimney is equally bad.

Always check the dryer for level position on the floor, after working on it or installing a vent system. Its operation is affected if it is wobbly or at a slant. Some legs are adjustable, others may have to be shimmed with small, thin pieces of wood.

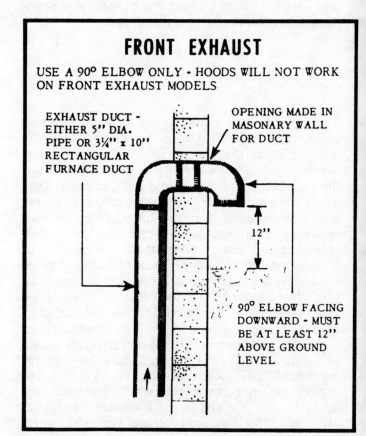

FRONT EXHAUST
USE A 90° ELBOW ONLY - HOODS WILL NOT WORK ON FRONT EXHAUST MODELS

EXHAUST DUCT - EITHER 5" DIA. PIPE OR 3¼" x 10" RECTANGULAR FURNACE DUCT

OPENING MADE IN MASONARY WALL FOR DUCT

12"

90° ELBOW FACING DOWNWARD - MUST BE AT LEAST 12" ABOVE GROUND LEVEL

Figure 116

EARLY MODELS
SENSITRON CHECK - OUT CHART
GAS DRYER MODELS

IMPORTANT – *Some early production dryers were not equipped with isolating transformers, Part No. 10-1342. On these dryers, check for correct polarity and proper grounding. In addition all settings in CONTROL SETTING column must be made to completely check out the SENSITRON automatic dryness control.*

CONTROL SETTING	OPERATION	CHECK AS FOLLOWS
Set TEMPERATURE control to "Hi" position and set TIME DRY cycle knob to 30 minutes.	Dryer in operation: – Lights on – Drum revolving – Main burner on **NOTE:** *Steps 1 through 4 can be eliminated if dryer is in operation. However, Steps 5 and 6 must be checked.*	1. Check source of power to dryer. 2. Place Neon test light from terminal "L1" of SENSITRON control to wire nut connection which holds gray service cord and black wire lead. Light indicates open drum protectomatic. Manually reset control and determine cause. 3. Place Neon test light across terminal "L1" and "H" of SENSITRON control. Also check across "L1" and "M". Light in either test indicates open contacts. Replace SENSITRON control. 4. Check other components which could cause failure of dryer operation such as: motor, motor throw-out switch, oven protectomatic, thermostat, door switch, solenoid valve, power pack, spark electrodes, magnetic vibrator, and lights. Also check for wire leads off, and broken or pinched wires. 5. Remove lead from SENSITRON terminal "TM". Check with Neon test light from removed lead to "W" terminal on SENSITRON control. Light indicates timer motor O.K. No light replaces SENSITRON control. 6. Place Neon test light across SENSITRON terminals "W" and "TM". Light indicates open in timer contact circuit. Replace SENSITRON control.
FOR DRYERS WITH ISOLATING TRANSFORMERS: Place grounding clip (taped to back of dryer) across sensing plate to baffle. If clip is not available, place a strip of metal from sensing plate to baffle to complete the circuit. Set SENSITRON DRY-DAMPNESS control to "Dry" and AUTOMATIC DRY cycle knob to "Normal"	Dryer in operation (Check-Out Step 7 must be made).	7. Place Neon test light between brown wire leads on transformer. (On some models it is necessary to remove the wire nuts.) Light indicates transformer O.K. No light, replace transformer. Remove clip after test.

"SENSITRON" CHECK-OUT CHART (CONT.)

CONTROL SETTING	OPERATION	CHECK AS FOLLOWS
Set TEMPERATURE control to "Air" and set *SENSITRON* DRY-DAMPNESS control to "Max Damp." Rotate AUTOMATIC DRY cycle knob through "Wool" mark. Set TEMPERATURE control to "Air" and set *SENSITRON* DRY-DAMPNESS control to "Dry." Rotate AUTOMATIC DRY cycle through "Wool" mark. Place grounding clip (taped to back of dryer) across sensing plate to baffle. If clip is not available, place a strip of metal from sensing plate to baffle to complete the circuit. Set TEMPERATURE control to "Hi" position and *SENSITRON* DRY-DAMPNESS control to "Max Damp". Set AUTOMATIC DRY cycle control knob to "Heavy"	An audible click should be heard and dryer should operate from 0 to 10 seconds. An audible click should be heard and the dryer should operate from 30 to 90 seconds. Dryer in operation: – Lights on – Drum revolving – Main burner on **NOTE:** *Despite the fact that dryer appears to be working satisfactorily, Step 9 must be made.*	8. If the audible click is not heard replace *SENSITRON* control. If click is heard but dryer fails to stop, remove terminal lead "7" or "8" (depending on *SENSITRON* involved — see note below) from *SENSITRON* control. If dryer then shuts itself off in the "Wool" setting, locate and correct shorted condition in the baffles, contact assembly, or slip ring assembly. If dryer continues to operate with lead removed, replace *SENSITRON* control. 9. Place Neon test light across terminals "7" or "8" (depending on *SENSITRON* involved — see note below) and "G" on *SENSITRON* assembly. Light indicates open in contact assembly, slip ring assembly, or sensing plates. Determine cause and correct.

IMPORTANT

Three styles of SENSITRON controls were used. The controls are identical in appearance but differ in internal wiring. As the result of these wiring changes, terminals 7 and 8 are reversed thus affecting checking procedures. Refer to Chart below for details:

Hamilton Part No.	Supplier Part No.	Identification Marking On Top Of Control Case	Checking Terminal To Be Used
10-1310	551-001	None	8
10-1341	551-001-1	GAS	8
10-1347	551-002-1	GAS 2	7

SECTION 4

SERVICE PROCEDURE
GAS COMPONENTS

GAS BURNERS, Later Models

The control of burner is made by one of two ignition systems:

A.   The direct action solid state spark ignition, *Figure 117*

B.   The mercury element standing pilot ignition, *Figure 118.*

Both systems consist of three major components:

1.   **Gas valve and regulator assembly**
2.   **The ignition assembly**
3.   **The burner assembly**

The complete gas control assembly, regardless of ignition system involved, is attached to the dryer base and is accessible from the front of the dryer. Remove the lower front panel when servicing or replacing any of the components. To remove the complete control assembly, follow these steps:

1.   Disconnect electrical power source.

2.   Shut off gas supply at main shut-off valve and disconnect fiber strip shield from shut-off valve union.

3.   Disconnect shut-off valve union nut.

4.   Remove screws securing assembly to base.

5.   Disconnect leads and lift out assembly.

GAS VALVE AND REGULATOR ASSEMBLY

The valve and regulator assembly control the flow of gas to the burner by means of a solenoid coil and regulates the gas pressure to the main burner and the pilot burner. (When L.P. gas is used, the regulator in the dryer is not used.)

PRESSURE REGULATOR

Burners are equipped and the pressure regulator adjusted at the factory for natural gas. The pressure regulator is pre-set to provide 3.5 inches of water column pressure at the main burner. If an adjustment is necessary, a manometer must be used to check and adjust the regulator.

SOLENOID COIL

The main solenoid coil is energized by a circuit controlled by the timer, thermostats and motor operation. This allows gas to flow to the main burner.

To remove the coils:

1.   Disconnect wires from coil.

2.   a.   On automatic spark ignition:

Remove screws fastening safety pack to valve and lift safety pack up.

b.   On mercury vapor pilot ignition:

Remove screws securing hold down bracket and lift off.

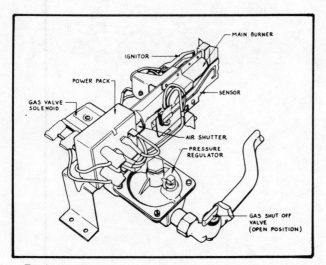

Figure 117. Direct Action, Solid State Spark Ignition Gas Burner

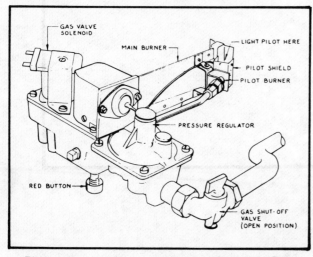

Figure 118. - Mercury Element Standing Pilot Ignition Gas Burner

3. Remove coil and install replacement.

NOTE: The coils, though appearing similar, are not interchangeable. The coil on solid-state is a D.C. coil while the coil on mercury vapor is A.C.

MAIN BURNER ORIFICE

The main burner orifice is attached to a fitting on the valve assembly. It regulates the amount of gas flowing to the main burner for the type of gas being used.

To change the main burner orifice, proceed as follows:

1. Shut off gas supply at main shut-off valve.

2. Remove the complete gas control assembly.

3. Remove rear screw securing left side mounting bracket to burner.

4. Disconnect pilot line tubing at filter end (mercury-vapor ignition only).

5. Remove burner from valve assembly.

6. Remove orifice from fitting.

Reassemble in reverse order.

IGNITION SYSTEMS
THE SOLID STATE SPARK IGNITION SYSTEM

This system provides a means of directly lighting the gas in the burner without a pilot light as an igniter. See *Figure 120.*

This system consists of:

1. **The power pack**
2. **The ignitor assembly**
3. **The heat sensor assembly**

THE POWER PACK ASSEMBLY, *Figure 119*

The power pack component, located on the gas valve and regulator assembly, controls the electrical circuits to the gas valve and to the igniter.

THE IGNITER ASSEMBLY, *Figure 120*

The igniter consists of a solenoid coil and spark electrodes (magnetic vibrator) and bracket, and is attached to the burner by two screws. Its function is

to produce the spark to ignite the gas at the burner.

Do not bend or tamper with the igniter arms as this will affect the spark given off by the electrodes. Also do not remove the coil and electrode from its mounting bracket. Evidence of disassembly will void warranty.

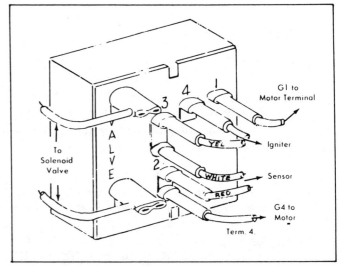

Figure 119- Control Panel Terminals

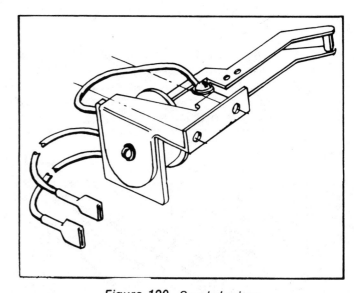

Figure 120.· Spark Igniter

THE HEAT SENSOR ASSEMBLY, *Figure 121*

The flame sensor assembly consists of a switch mounted on a sensor bar. It is attached to the burner with screws and a bracket. Once the burner gas is ignited, the flame impinges on the sensor bar causing its arm to expand. This opens the switch, thereby opening the igniter circuit and stopping the sparking action.

When the burner flame is extinguished, the sensor bar arm contracts, thus closing the switch and permitting the igniter to again spark.

OPERATION OF DIRECT IGNITION COMPONENTS, *Figure 122.*

The solid state spark ignition operates as follows:

1. The circuit in the dryer is completed from the momentary contact switch, through the timer, the thermostats, and the centrifugal switch in the motor.

2. The circuit to the power pack is made from terminals 1 and 4 on the motor to terminals 1 and 2 on the control panel of the power pack.

3. Both the circuits for the 110 volt AC magnetic vibrator and the rectified 60 volt DC gas valve solenoid are energized. See *Figure 122* for energized circuits to start the burner. The gas valve opens and the vibrator sparks to light the gas.

4. The flame from the burner heats the sensor bar. The bar expands and opens the switch on the bar to break the circuit to the spark igniter to stop the sparking, *Figure 123* for circuits after igniter is off.

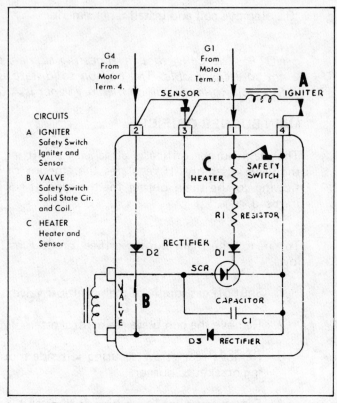

Figure 122 Closed Circuits to Ignite the Burner

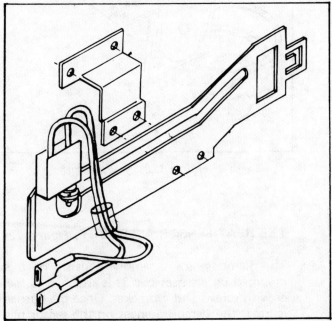

Figure 121 - Heat Sensor Assembly

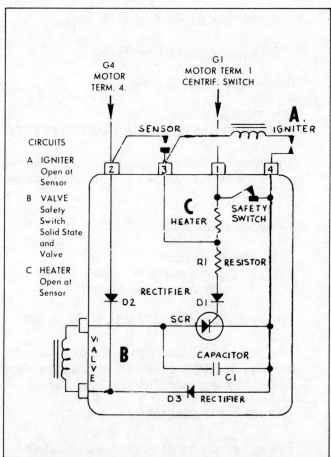

Figure 123 - Closed Circuit While Burner Is Ignited

The dryer will operate for the balance of the cycle with the drying chamber temperature being controlled by the thermostat. As the thermostat cycles, electrical power to the gas controls is interrupted. This shuts off the gas supply, thus extinguishing the burner flame. The dryer continues to operate without heat until the thermostat contacts close, again permitting electrical circuitry to the gas components. Ignition operation is also affected by:

a. Cycling of the safety thermostat.

b. Stoppage of motor, such as would be the case by opening the door.

c. Shutting off of timer.

d. Interruption of gas supply.

5. To prevent the flow of gas if ignition does not take place, a warp safety switch is in the circuit to the gas valve and the vibrator. A heater is also in series with the heat sensor switch and is attached to the warp switch. If the sensor switch does not break the circuit to the spark igniter in thirty to seventy seconds, the heater warms the warp switch to break the circuits to the valve and igniter, *Figure 124.*

The warp switch heater will continue to heat until the circuit is broken to the power pack. Approximately five minutes should be allowed for the warp switch to cool and return to its closed position before turning the time back on.

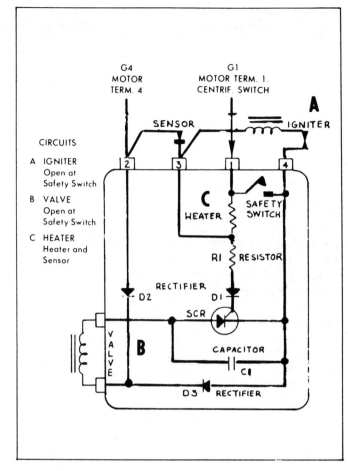

Figure 124. - *Circuit When Burner Does Not Ignite*

SOLID STATE CHECK-OUT PROCEDURE

Operation	Indicator You Look Or Listen For	If Indicator Is Not Seen Or Heard, Check As Follows:

Because this check-out procedure is for components only, it will be necessary to check for "wire off", "loose terminal", or "open wire" conditions as necessary whenever test light check result is "NO LIGHT". (Extension leads on test light may be useful. Use light rated no lower than 60 watts.)

| With timer in off position, open loading door. | Light from drum and ozone lights. (Lights not supplied on some models; then steps 1, 2 and 3 must be checked.) | 1. **POWER FAILURE:** Place neon test light from S, L1 terminal on timer, depending on model involved, to common junction of white wires on timer. NO LIGHT indicates no power at receptacle, poor connection of quick disconnects, or poor connection at multiple connector block (some models). Visually check and correct condition. |
| | | 2. **DOOR SWITCH (CONTACTS 1 AND 2):** Place neon test light from E terminal on timer, to common junction of white wires on |

Operation	Indicator You Look Or Listen For	If Indicator Is Not Seen Or Heard, Check As Follows:
		timer. LIGHT indicates door switch contacts closed. NO LIGHT, remove switch and check continuity through contacts. Replace switch if defective.
		3. **FLOOD LIGHT AND OZONE LIGHT:** Disconnect the wiring harness leads from ozone light and then place neon test light across the disconnected harness leads. NO LIGHT indicates flood lamp burned out. LIGHT indicates flood lamp OK. Replace ozone bulb.
Turn timer to 5-minute position. (On non-adjustable thermostat models, set timer at 5-minute position.) Turn thermostat to air fluff. Close the loading door.	Dryer motor runs and drum revolves.	4. **TIMER:** Place a neon test light across timer contacts S, and E. LIGHT indicates contacts open. 5. **DOOR SWITCH (TERM NC TO C):** Place neon test lamp from common junction of blue wires on timer to common junction of white wires on timer. LIGHT indicates door switch contacts OK. (Or place test light across contacts NC to C, LIGHT indicates switch needs adjustment in cabinet or replacement.) If NO LIGHT, check plunger travel in relation to closed door. If OK, replace switch. 6. **MOTOR:** Place neon test lamp across motor terminals 4 and 5. LIGHT indicates power to motor. Trouble may be in motor or motor switch. Repair or replace as necessary.
Turn manual gas valve to off, thermostat to heat position and set timer at 30-minute position.	Spark at magnetic vibrator.	Place neon test light across terminals 1 and 2 of solid-state power pack. LIGHT indicates steps 7, 8, 9, and 10 are OK. Proceed to step 11. NO LIGHT, check as follows: 7. **TIMER:** Place a neon test light across terminal A and S of timer. LIGHT indicates contacts open. Replace timer.
Turn manual gas valve to off, thermostat to heat position and set timer at 30-minute position. (On dryers without timer cycle set on dry setting.	Spark at magnetic vibrator.	8. **THERMOSTAT:** Place a neon test light across terminal 1 and 2 of control thermostat. LIGHT indicates contacts open. Replace thermostat. 9. **SAFETY THERMOSTAT:** Place neon test light across terminals 1 and 3 of a 3 terminal thermostat or 1 and 2 of a 2 terminal thermostat. Light indicates open to burner. Determine cause and if defective, replace.
Turn manual gas valve to off, thermostat to heat position and set timer at 30-minute position. (On dryers without timer cycle set on dry setting.	Spark at magnetic vibrator	10. **MOTOR THROW-OUT SWITCH:** Turn timer off and disconnect red lead from valve safety pack. Turn timer on. Place neon test lamp from disconnected lead to terminal 1 of power pack. NO LIGHT indicates open mo-

Operation	Indicator You Look Or Listen For	If Indicator Is Not Seen Or Heard, Check As Follows:
		tor throw-out switch. Remove motor and repair or replace switch. Reconnect red lead to terminal 1.
When making checks 11 through 15, remember that warp switch is in circuit and it will open in thirty to seventy-five seconds. When this occurs, turn timer off for five minutes to allow switch contacts to reset.		11. FLAME DETECTOR SWITCH: Place neon test lamp from terminal 1 to terminal 3 of power pack. LIGHT indicates sensor switch OK. NO LIGHT — replace.
		12. WRAP SWITCH: Place neon test light across terminals 1 and 4 of power pack. LIGHT indicates warp switch contacts are OPEN. Determine cause. See note on warp switch contact opening, at left. If defect lies in power pack, replace.
		13. MAGNETIC VIBRATOR: Since checks 12 and 13 indicate you have electric power at vibrator, failure for vibrator to work indicates component is defective. Check for open in lead wires or dirt particles on contact points. Failure can be verified by attaching a 115 volt test service cord directly to removed vibrator leads. Replace magnetic vibrator if necessary.
		14. POWER PACK ASSEMBLY: Place neon test light across terminals 3 and 4 of power pack. LIGHT indicates power pack OK. NO LIGHT — replace.
Turn manual gas valve to on position. Leave timer and thermostat as set in last operation.	Gas solenoid energized and main burner ignited.	15. Place neon test light across blue wires on solenoid coil. LIGHT indicates power to coil. If coil will not energize to open valve, replace. This is a D.C. circuit. Do **not** test coil with 115 volt service cord.

MERCURY ELEMENT STANDING PILOT IGNITION

This system provides:

1. A constant flow of filtered gas to the pilot burner.

2. A safety shut off if the pilot light goes out.

3. Controls the gas supply line up to the main solenoid valve.

The system consists of: *(Figure 125)*

1. Plunger assembly (Item 2)
2. Plate and latch assembly (Item 3)
3. Mercury element (Item 4)
4. Pilot burner assembly (Item 5)
5. Pilot orifice (Item 6)
6. Pilot filter (Item 7)

OPERATION

PLUNGER

The plunger assembly controls the flow of gas to the pilot burner and to the main solenoid valve.

1. With the button out and pilot not lit, no gas to pilot or main solenoid.
2. With button in, gas is open to the pilot burner, but not to the main solenoid.
3. With button out and pilot lit, gas is open to the main valve.

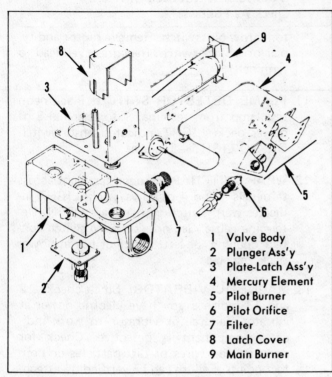

1	Valve Body
2	Plunger Ass'y
3	Plate-Latch Ass'y
4	Mercury Element
5	Pilot Burner
6	Pilot Orifice
7	Filter
8	Latch Cover
9	Main Burner

Figure 125. - *Mercury Element Pilot Ignition System*

LATCHING MECHANISM AND MERCURY ELEMENT

By pressing in on the plunger assembly, it pushes in on a spring-loaded shaft with attached locking disc mounted to valve plate, *Figure 127.* Also fastened to the valve plate is a latch arm and pin. With the plunger pressed in, the disc is pushed up above the height of the latch pin. To activate the latch arm and move the pin, a mercury element assembly consisting of bulb, capillary and diaphragm connects the pilot burner to the latch arm and pin. The bulb is mounted on the pilot burner holder and the diaphragm is fastened to the mounting plate by the latch arm.

When cold, the mercury exerts no pressure on the diaphragm. The latch pin is free from the disc. When sufficient heat is applied to the mercury, it exerts pressure on the diaphragm and the latching mechanism, causing the latch pin to move under the locking disc. Only when the plunger is depressed will the latch pin catch under the disc to hold the pilot gas line open. About ninety seconds is necessary for proper latching. The plunger may then be released. The gas line is now open to the solenoid valve.

If for any reason the pilot burner goes out, the mercury in the element cools and releases the pressure under the latch operating diaphragm, which releases the mechanism so that the spring on the flow valve

returns to its normally closed position. This action stops the gas flow to the pilot burner and the main gas valve *Figure 127* and *Figure 128.*

To check latch mechanism action:

1. Remove the cover.
2. In the cold position there should be .017" clearance + .010 between the disc and the latch pin.
3. To adjust the latch pin use a 1/32" allen wrench.
4. Check action by lighting pilot. The latch pin should move under the disc within ninety seconds.
5. If pilot burner does not remain on after lighting, check system as follows:
   a. Check to make sure orifice is not partially plugged or defective.
   b. Check for bent or damaged mercury element or brackets.
   c. If element is in correct position but gas flame does not impinge on it, check for blockage in either pilot burner or orifice.
   d. Check latch arm mechanism for proper action.

NOTE: When installing a new element, the correct clearance, .017" + .010, must be made between the disc and latch pin.

100% GAS SHUT OFF SYSTEM

Some pilot models have a pilot safety shut off. These models have an additional, normally-open, safety thermostat mounted in the heat chamber, and a safety solenoid mounted by the mercury element latching mechanism.

If the temperature rises above 210° F in the drum and neither the cycling thermostat nor the safety thermostat open, the second safety thermostat closes the circuit to the safety solenoid which push-

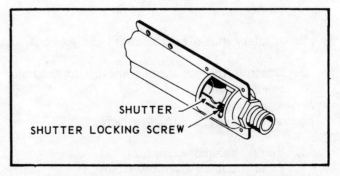

SHUTTER

SHUTTER LOCKING SCREW

Figure 126 - *Air Shutter Adjustment*

es the latch pin from the disc and shuts off the gas supply to the pilot light. The solenoid buzzes to warn the operator that the gas supply has been shut off, and the dryer should be stopped.

PILOT ORIFICE

The pilot orifice controls the amount of gas flowing to the pilot burner for the type of gas being used.

To change the pilot orifice, proceed as follows:

1. Shut off gas supply at main shut-off valve.
2. Remove the complete burner assembly.
3. Disconnect pilot line tubing at filter end.
4. Remove tubing at orifice end.
5. Use one wrench to hold the pilot body while removing the orifice with another wrench. This procedure must be followed to avoid undue stress on components.

Reassemble in reverse order.

PILOT FILTER

The pilot filter is part of the valve body and filters the gas going to the pilot to prevent the pilot orifice from becoming plugged by dust or foreign particles. The pilot filter can be serviced as follows:

1. Shut off gas supply at shut-off valve.
2. Disconnect 1/4 inch tubing from filter and pilot orifice by removing compression nuts.
3. Remove filter from valve. The filter is part of the valve. When these are involved, remove the three screws holding pilot cover to valve and remove gasket, screen, and filter material.
4. Replace filter material and reassemble in reverse order.

BURNER ASSEMBLY

The burner, a universal in-shot type burner, consists of the burner tube, air shutter and flame deflector.

AIR SHUTTER ADJUSTMENT, *Figure 126*

The primary air shutter adjustment is located at the base of the burner. The shutter adjustment should be set for a soft blue flame. Too sharp a flame, especially if L.P. gas is used in burning, may result in

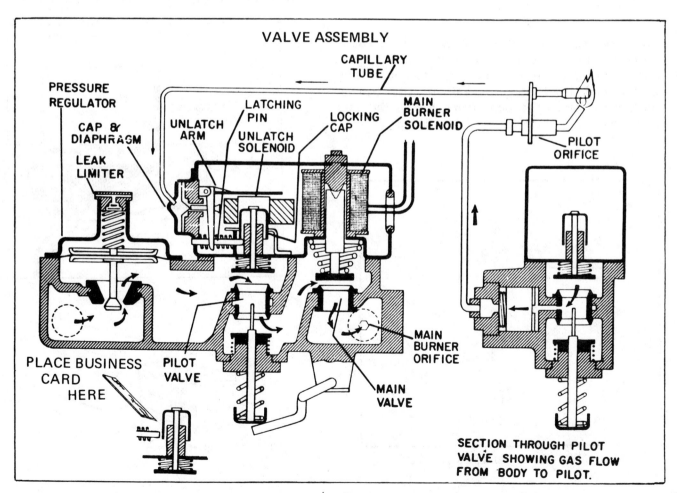

Figure 127- Valve Assembly

flame sensor or igniter point damage and subsequent erratic operation.

NOTE: The unlatch mechanism can be adjusted by turning adjusting screw indicated, Figure 127. The thickness of an ordinary business card should exist between the heavy part of plunger and latch pin when mercury tube is in a cold state.

PILOT SAFETY DEVICES

A cut-away view of the complete valve assembly, *Figure 127,* can better illustrate the pilot safety devices. Assuming the pilot becomes extinguished, the pressure on the diaphragm will be removed and the spring loaded latching pin will be retracted from under the locking cap. The pilot valve plunger, which is spring loaded, closes off the valve port to the pilot, thus stopping all flow of gas.

When gas valve is open and air bled from gas supply lines, the pilot is ready to be lighted. The gas has flowed up to the pilot valve. The latching pin is not engaging the locking cap.

The pilot reset lever has been moved to the vertical position. This closes the bottom of the pilot valve and opens the top part, allowing gas to flow to the pilot burner. The locking cap was also raised by the pilot reset lever. The pilot flame has caused the latching pin to be pushed under the locking cap.

The pilot reset lever has been moved back to its normal position, opening the bottom of the pilot valve. The upper part of the pilot valve is held open by the latching pin under the locking cap. When the main burner solenoid is energized, it lifts the main burner valve plunger and lets gas flow to the main burner.

GAS CONVERSION KITS

The dryers are assembled for use with natural gas. Kits are available to convert from natural gas to L.P. gas, or if need be, to convert from L.P. gas to natural gas.

NOTE: When converting spark ignition dryers, the flame sensor must also be changed for proper operation of the burner.

CHANGE	ORDER KITS	FLAME SENSOR
Spark Ignition Nat. to L.P.	F76885-1	Furnished in kit
Standing Pilot Nat. to L.P.	F76885	None used
Spark Ignition L.P. to Nat.	F76886	F76888-1
Standing Pilot L.P. to Nat.	F76886	None used

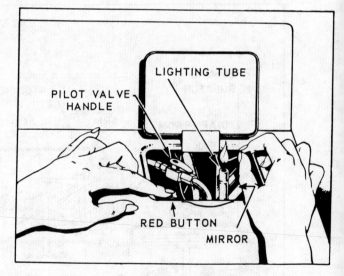

Lighting The Pilot

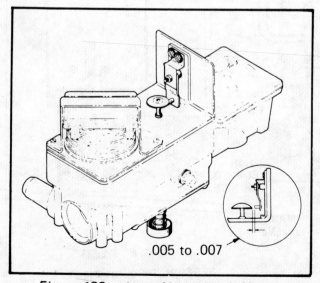

Figure 128 - Later Model Latch Mech.

.005 to .007

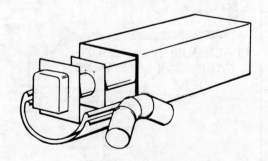

Vent Kit

LATE MODELS, SENSITRON DRYNESS CONTROLS CHECK CHART

Details on setting of the Variable Temperature Selector are included in this chart because the settings and subsequent operation is directly associated with the operation of the sensitron.

Cycle	Variable Temperature Selector (Left Knob)	Sensitron Dryness Selector (Right Knob)
		Each cycle provides a range of dryness. For greater dryness within the cycle, turn the knob clockwise; for greater dampness turn the knob counterclockwise. When the dryer is in operation the knob rotates counterclockwise until the indicator reaches the "OFF" position.
DAMP MEDIUM DAMP DRY	**Setting:** Operator turns knob until indicator points to LOW, MEDIUM, WOOL, PERMANENT PRESS or REGULAR setting; depending on garments being dried. **Function:** Controls temperature in drying chamber. When degree of heat called for is reached, control interrupts heat circuit. Circuit remains open until temperature lowers.	**Setting:** Operator turns knob to either DAMP, MEDIUM DRY or DRY setting. The setting depends on type and size of load being dried and degree of dryness desired. For complete dryness the control would be set in DRY; if it were to be partially dried, the setting would be DAMP to MEDIUM DAMP, depending on dampness desired. A smaller load requires lower settings. All settings are followed by a short DE-WRINKLE period. **Function:** Controls the length of operation by measuring the moisture content of clothing being dried. When a predetermined degree of dryness is attained, the dryness control starts the time cycle, the length of which depends on the setting chosen.
WOOL	**Setting:** Operator turns knob until indicator points to WOOL setting. **Function:** Controls temperature in drying chamber. When degree of heat called for is reached, control interrupts heat circuit. Circuit remains open until temperature lowers.	**Setting:** Operator turns knob to WOOL setting. The dryness control operates the same as above except there is no DE-WRINKLE or cooldown period. **Function:** Controls the length of operation by measuring the moisture content of clothing being dried. When a predetermined degree of dryness is attained, the dryness control starts the time cycle, the length of which depends on the setting chosen.
AIR FLUFF	**Setting:** Operator turns knob until indicator points to AIR FLUFF setting. **Function:** The control mechanically holds the heat circuit open, providing a no heat cycle.	**Setting:** Operator turns knob until indicator points to AIR FLUFF setting. The dryness control operates the same as above except there is no heat during entire cycle. **Function:** Controls the length of operation by measuring the moisture content of clothing being dried and then immediately starts the time cycle, the length of which depends on the setting chosen.

LATE MODELS, SENSITRON DRYNESS CONTROLS CHECK CHART

NOTE: All settings in **CONTROL SETTING** column must be made to completely check out dryness control. All that is required for testing is a neon test light (rated at not less than 60 volts). When the test calls for a **LIGHT**, both elements should glow, and when it calls for a **HALF LIGHT**, only one element should glow.

Control Setting	Operation	No Operation Check As Follows:
Set thermostat (VARIABLE TEMPERATURE control) to "REGULAR" and set Sensitron Dryness Selector (timer) knob in "WOOL" cycle.	Dryer in operation: — Light on — Drum revolving — Burner on **NOTE:** *Steps 1 and 2 can be eliminated if dryer is in operation. However, steps 3 and 4 must be checked.*	1. Check source of power to dryer. 2. Place neon test light from terminal "L1" of timer to common junction of white wires. LIGHT indicates power at timer. NO LIGHT indicates open in service cord, molex connection or wiring harness. Check and correct as necessary. 3. Place neon test light across timer terminals "L1" and "H", and also across "L1" and "M". A LIGHT in either test indicates open contacts. Replace timer. 4. Place neon test light across terminals 1 and 3 on moisture control. LIGHT indicates power to moisture control. Proceed to Step 6. NO LIGHT — check Step 5. 5. Use Mercury-Vapor or Solid State ignition check-out charts to check other components which could cause NO LIGHT such as: thermostat, heat chamber and fan Hi-Limit protectomatics, gas ignition controls, motor, motor throw-out switch and door switch. Also check for leads off, broken or pinched wires.
Cycle settings same as above. Place grounding clip (any piece of metal) across sensing rod to baffle to complete circuit.	Dryer in operation: — Light on — Drum revolving — Burner on **NOTE:** *Despite the fact that the dryer appears to be working satisfactorily. Steps 6, 7 and 8 must be made.*	6. Disconnect lead from terminal "2" on moisture control. Place test light from terminal "2" to terminal "TM" on timer. HALF LIGHT indicates dryness control OK. FULL LIGHT, replace dryness control. Re-attach disconnected 7. Place lead neon test light from terminal "5" on dryness control to controls mounting bracket. NO LIGHT indicates circuit, OK; HALF LIGHT indicates open in contact assembly, slip ring assembly or sensing rods. Determine cause and correct. 8. Remove tan lead from terminal 5. Place test light from removed lead to terminal 5. HALF LIGHT — OK. NO LIGHT indicates open in wiring, sensing rods, contact and slip ring assemblies. Determine cause and correct. Re-attach disconnected lead.
Cycle settings same as above. Remove grounding clip.	Dryer in operation: — Light on — Drum revolving — Burner on	9. Remove timer motor lead from terminal "TM". Check from removed lead to terminal "TM". FULL LIGHT — timer motor OK. NO LIGHT — replace timer. Re-attach disconnected lead.

MANUAL IGNITION

EARLY MODELS

The gas dryer has had a great selection of ignition methods, both manual and automatic. In the early manual controls, the operator must relight the pilot if it is shut off by any of the safety thermostats or controls. Relighting the pilot is illustrated in *Figure 129.*

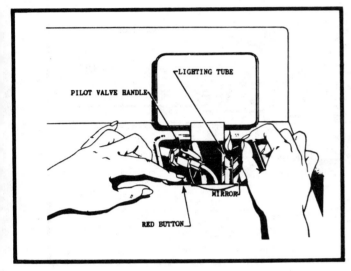

Figure 129

In the later models, manual ignition controls shut off the main burner line of gas supply only. The pilot continues to burn if the protectomatic has cycled due to restricted air circulation. The differences in the circuits will be explained in the individual instructions for the various systems.

TO LIGHT THE PILOT BURNER.

An access door is provided to give exposure to the pilot valve, the pilot valve shut off handle, the reset button and the lighting tube. In some models the reset button (red) is replaced by the baso handle (lever). The button is depressed to operate, the lever is lifted. After opening the pilot valve, the button is depressed (lever lifted) and a match applied to the lighting tube opening. There may be slight "pop" as the pilot lights. In most of the models, a mirror (*Figure 129*) is supplied to observe the flame. After the flame is seen to appear, the button (lever) is kept on for about 40 seconds, at which time the pilot light should stay on. If it goes out, repeat the process once more.

Shut the pilot light off by turning the handle to "off" position. If the pilot goes out for any reason, always wait for five minutes to allow any trapped gas to be dispelled before relighting.

The pilot lighting tube should be properly aligned to insure easy lighting of ignition. *Figures 130 and 131* illustrates both the top and side view of the pilot and main burner assembly as it should be aligned. Any diviation should be corrected.

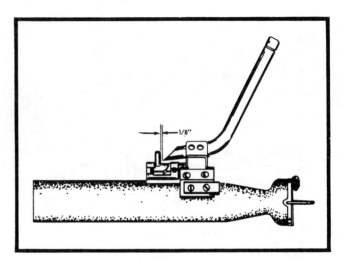

Figure 130

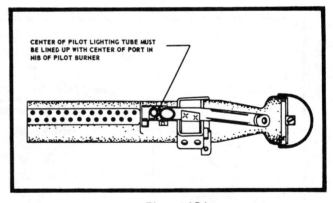

Figure 131

A typical gas flow system is shown in *Figure 132* which may be slightly different in some cases but will chart the relations of the manual ignition generally used. A brief description of its operation follows:

From the main gas supply line the gas will travel though the shut-off valve to the pressure regulator then to the baso valve.

If the pilot light is burning and the thermocouple fastened to the pilot burner is heating due to the flame impinging on it, there will be a low voltage current produced in the thermocouple assembly. This current will activate the electro-magnet in the baso valve and the valve will be held open allowing gas to pass through to the burner. If the pilot light goes out, the current will not be generated and the baso valve closes the gas supply valve. How the millivolt circuit is checked for correct operation is

shown in *Figures 133, 134 and 135*, which both the oven protectomatic types and those without protectomatic control are described with specific checks for millivolt readings to determine the condition of the valve and the thermocouple circuit. A millivolt meter must be used.

BASO VALVE.
There are three sections to the baso valve - the

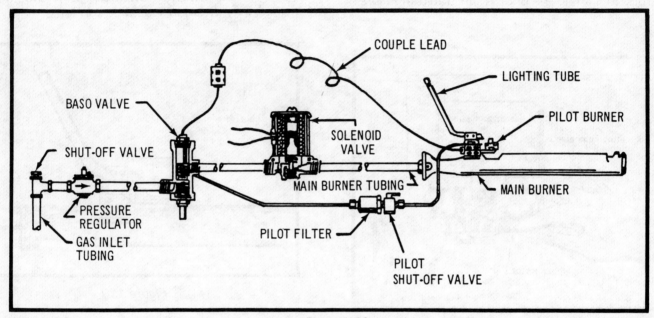

Figure 132

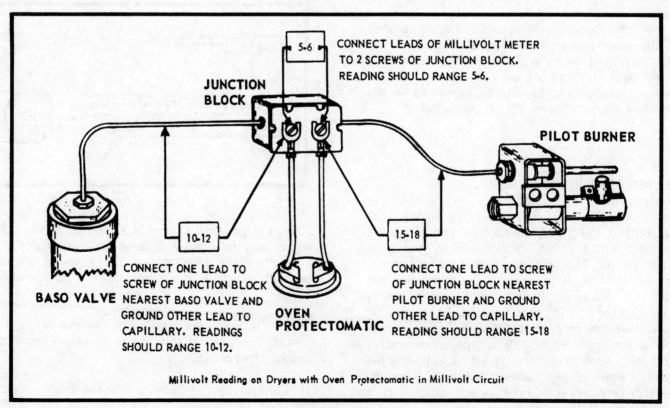

Millivolt Reading on Dryers with Oven Protectomatic in Millivolt Circuit

Figure 133

electra-magnet, the plunger valve and the thermo-couple. The plunger valve (red button or lever) is the one that allows gas to flow to the pilot burner when manually activated. The electromagnet, when energized, opens the valve to allow gas to flow to the solenoid valve. At the same time it allows gas to flow to the pilot when the red button or lever is released. When the gas pressure drops too low or the safety thermostat opens the millivolt circuit (the thermocouple cools) the base will snap shut and cut off the gas supply to the dryer.

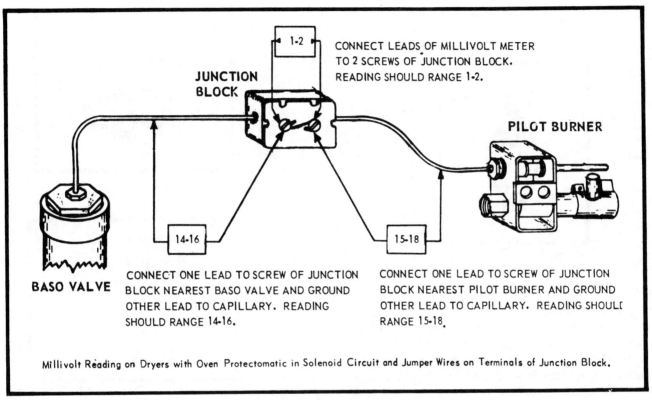

JUNCTION BLOCK

1-2

CONNECT LEADS OF MILLIVOLT METER TO 2 SCREWS OF JUNCTION BLOCK. READING SHOULD RANGE 1-2.

PILOT BURNER

BASO VALVE

14-16

CONNECT ONE LEAD TO SCREW OF JUNCTION BLOCK NEAREST BASO VALVE AND GROUND OTHER LEAD TO CAPILLARY. READING SHOULD RANGE 14-16.

15-18

CONNECT ONE LEAD TO SCREW OF JUNCTION BLOCK NEAREST PILOT BURNER AND GROUND OTHER LEAD TO CAPILLARY. READING SHOULD RANGE 15-18.

Millivolt Reading on Dryers with Oven Protectomatic in Solenoid Circuit and Jumper Wires on Terminals of Junction Block.

Figure 134

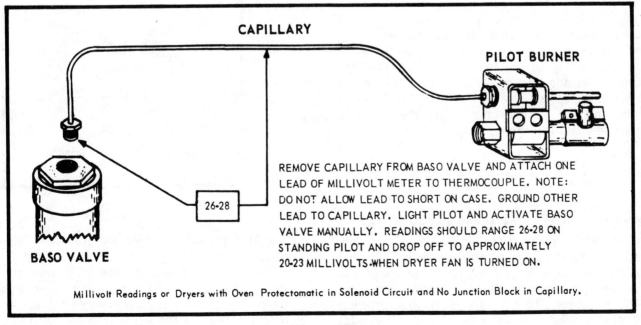

CAPILLARY

PILOT BURNER

26-28

REMOVE CAPILLARY FROM BASO VALVE AND ATTACH ONE LEAD OF MILLIVOLT METER TO THERMOCOUPLE. NOTE: DO NOT ALLOW LEAD TO SHORT ON CASE. GROUND OTHER LEAD TO CAPILLARY. LIGHT PILOT AND ACTIVATE BASO VALVE MANUALLY. READINGS SHOULD RANGE 26-28 ON STANDING PILOT AND DROP OFF TO APPROXIMATELY 20-23 MILLIVOLTS WHEN DRYER FAN IS TURNED ON.

BASO VALVE

Millivolt Readings or Dryers with Oven Protectomatic in Solenoid Circuit and No Junction Block in Capillary.

Figure 135

There are times when the baso valve may require service. The hood assembly, *Figure 136,* can be replaced but the important thing to remember is the danger of stripping the threads if great care is not excercised. The best way to loosen the retaining nut is to strike the wrench with your hand in short sharp blows and NOT in a steady pressure pull. The threads in the body may be damaged and the whole unit will have to be replaced.

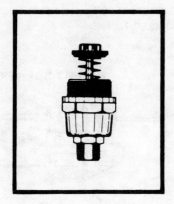

Figure 136

SOLENOID VALVE (Manual Ignition).
The solenoid valve *Figure 137* is a valve that is actuated by the electromagnet in the body which opens and closes the valve. The electromagnet may be replaced by removing the valve cover and detaching the wires from the circuit. Pull the wires out of the valve cover outlet and replace the unit. Reverse the steps to reassemble. Operation can be checked by removing the leads and attaching a 115 volt, 60 cycle AC current directly to the coil. A click is heard if the unit is serviceable. It may become necessary to clean the plunger assembly. This can be done by removing the coil from the shaft and takingout the four screws that hold the assembly to the casting. Wipe off with a clean cloth but never use abrasives of any kind. Be careful clamped in a vise. Pressure on the coil case can also damage it when tightening.

"REDI-MITE" MANUAL IGNITION SYSTEM
An improved version of the manual ignition system eliminates the baso valve, the single solenoid and the regulator by using a combination regulator and dual solenoid arrangement, *Figure 138.* Instead of a thermocouple there is an expansion rod that operates a pilot switch, *Figure 139.* The pilot

operates a pilot switch, *Figure 129.* The pilot solenoid coil operates on 115 volt circuit which replaces the millivolt system. Dryers equipped with this system are marked with an "R" in the serial number of the model rating plate.

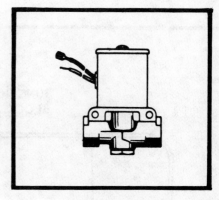

Figure 137

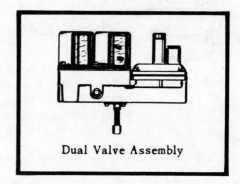

Dual Valve Assembly

Figure 138

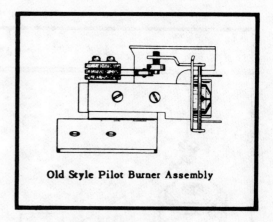

Old Style Pilot Burner Assembly

Figure 139

To light the pilot burner on "Redi-Mite" Ignition systems, the power supply plug must be plugged in at the power source. The pilot shut-off valve is turned to the open position and the lighting tube pulled up to its 45 degrees angle for the lighting position. Depress the red button and light the tube.

Continue to depress the red button until a distinct click is heard. That means that the expansion rod has heated and actuate the pilot switch arm, closed the circuit to the solenoid coil and that click is the solenoid coil being opened by the current. Release the button and the pilot should stay on. To turn the pilot off, simply turn off at the pilot shut-off valve.

"Rede-Mite" operation is simple. After you have lit the pilot, the timer is set for the length of operation desired. Contacts A and C are closed to B, the common terminal. Two things take place: The lights, timer motor and the drive motor are activated by th closing of contacts B and C. The closing of the contacts B and A energizes the main burner solenoid and the gas is allowed to flow to the main burner where it is ignited by the pilot light. A typical diagram using this burner assembly is shown in *Figure 140.*

If the loading door is opened during operation the door switch will break the circuit to the timer motor, thereby stopping timer action; also the motor which will stop the fan and drum. It also breaks the circuit to the main burner solenoid and the main heat source is shut off.

If the pilot goes out for any reason, during a timed cycle the expansion rod will cool and contract, thus causing the pilot switch to open the circuit to the pilot solenoid. This suts off the valve.

Because all of the gas that goes to the main burner solenoid must first pass through the pilot burner solenoid, it is a certainty that no gas can flow through either of the solenoids if the pilot solenoid is not energized. If there is any escape of gas a defective pilot solenoid is evident.

AUTOMATIC IGNITION

COMPONENT SYSTEM (Mc. Quay Norris)

One of the early automatic ignition systems to be used on the Hamilton dryer was the "Component" system. Because there is a development connection between this system and the "Auto-Mite", and still later, the "SIGI" systems, we shall start with the oldest method and compare them later, before turning out attention to the troubleshooting charts for each model. See *Figure 141*, sample schematic.

The timer is manually set for the desired period of operation. The contacts A and C are closed to the common terminal B and like the manual "Redi-Mite" system the B and C contacts energize the lights, the timer motor and the drive motor. In addition to this, the pilot solenoid (allowing gas to the pilot burner) and the step-down transformer (producing a 2½ volt secondary current) is energized to provide heat. After the gas is ignited by

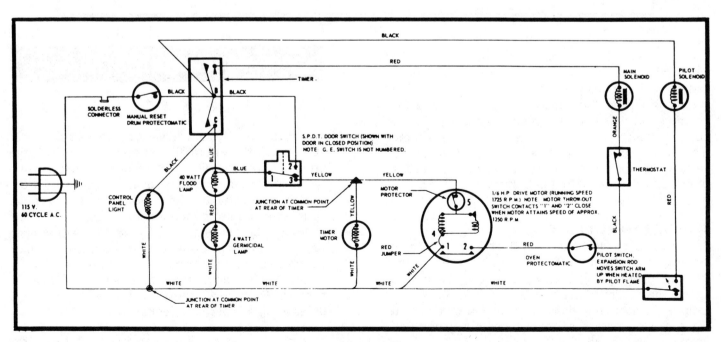

Figure 140

the glow coil, the stainless steel expansion rod is expanded by the heat of the flame. This closes the pilot switch contacts which energize the main burner solenoid circuit, allowing gas to the main burner where it is ignited by the pilot flame. At the same time, the control relay is started and so interrupts the transformer primary circuit, shutting off the flow coil. (On some models using LP gas, a warp switch heater is activated.)

In the cases where LP gas is used and the wrap switch used, the heater will open the contacts of th wrap switch assemblya nd the circuit to the transformer primary and the pilot solenoid is broken. This occurs if the main burner circuit is not energized within a period of 3 to 5 minutes. To re-energize the circuit, it will be necessary to turn the timer off and wait for the wrap switch to cool for about five minutes.

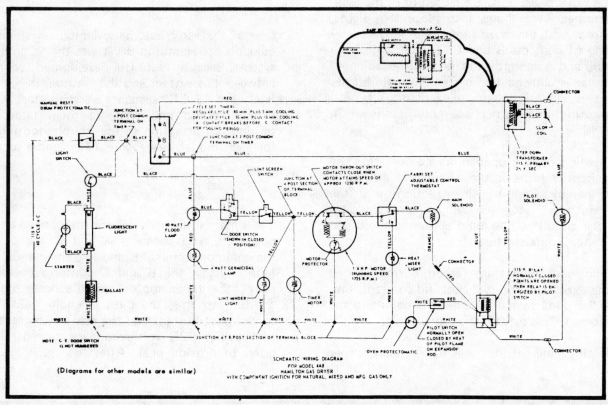

Figure 141

If the loading door is opened, the door switch will break the circuits to the timer motor, the drive motor and the main burner solenoid. If the pilot goes out during the timed cycle, the cooling expansion rod will open the pilot switch, the circuits to the main solenoid and the control relay coil. This latter action will cause the relay contacts to re-establish the transformer primary circuit and the glow coil will re-ignite the pilot burner.

FAILURE TO RE-LIGHT.

If it should fail to light again the gas will continue to flow in to the dryer for the timed cycle selected by the operator. The amount of gas escapage is below the maximum set by the American Gas Association and so is not dangerous. It will be detected by the customer as a no-heat complaint and will require a service call.

AUTO—MITE SYSTEM

The Auto-Mite system is very much like the older "Component" Automatic Ignition system, with the principle change being that there is no relay in the system. Instead, the step-down transformer is de-energized directly by the action of the single pole, double throw pilot switch incorporated into the new style pilot burner assembly, *Figure 142.* At the same time that the transformer circuit is broken, the pilot switch also completes the circuit to the main burner solenoid. This eliminates another source of possible trouble. *Figure 143* shows a typical diagram of a dryer using this burner assembly.

The timer is set for the desired length of time for

the operation and this closes the contact points of the timer to B and C, which activates the lights, the timer motor and the drive motor. The contacts B and A are also closed and the pilot solenoid allows gas to flow to the pilot burner as the step—down transformer primary furnishesthe 2½ volts secondary current to energized the glow coil and thus the gas is ignited in the pilot burner.

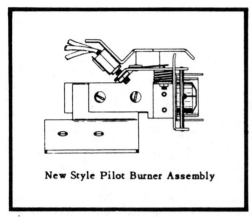

New Style Pilot Burner Assembly

Figure 142

After the gas is ignited in the pilot burner, the resultant flame will heat the stainless steel expansion rod and, as it expands, it will cut off the electrical circuit to the transforer primary and complete the gas circuit to the main burner, all by the action of the S.P.D.T. pilot switch.

As long as the dryer is in the normal selected time cycle it will continue operating. Should the loading door be opened, the door switch breaks the circuit to the timer motor, the drive motor and the main burner solenoid which shuts down the operation.

In case the pilot goes out during the timer cycle, the cooling of the expansion rod will reverse its original action, shut off the main burner gas supply and reactivate the circuit to the transformer primary. This will again cause the glow coil to re-ignite the pilot gas burner. Should the pilot fail to light for any reason, the pilot gas will coninue to flow a small amount of gas into the dryer until such time as timed cycle ends. This will require a service call to determine the reason for the no-heat condition.

All dryers with the LP gas supply will have a warp switch incorporated into the circuit. In this case, the heater of the warp switch will open the contact and break the circuit to the transformer primary. This occurs only after a period of 3 to 5 minutes of failure to ignite the main burner. To re-energize the circuit, the timer will have to be turned off to allow the heater time to cool. Wait at least 5 minutes for it to cool before attempting to re-energize.

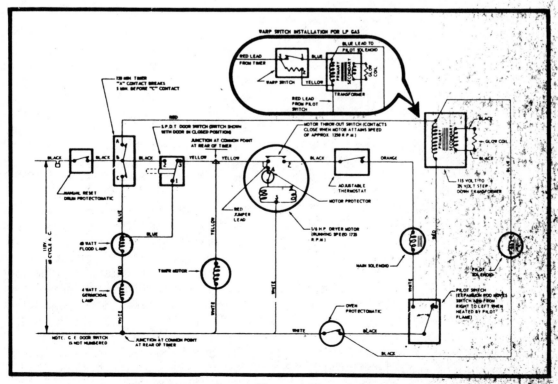

Figure 143

SIGI IGNITION

A pilotless type of ignition system, called the "SIGI" for short, has a sparking ignitor, applied directly the the main burner, and using 115 V current source to operate the ignitor.

To identify models that use this system, look for the letter "S" in the model number and also the letter "S" or "D" in the serial number on the rating plate. "S" in the serial number designates an early production model while "D" is current production.

The timer set for the desired length of time for of time for operation and the contacts A and C are closed to common terminal B. On those models having SENSITRON dryness control, the terminals closed are H and M to common terminal L1.

Closing of contacts B and C (M and L1) will energize the timer motor and the drive motor, as well as the lights.

Closing of the contacts A and B (H and L1) energize the magnetic vibrator (through a series circuit) — the normally closed flame detector switch, the normally closed warp switch, the thermostat, the oven protectomatic, and the motor throw-out switch. The magnetic vibrator furnishes sparks to ignite the gas at the burner.

At the same time, the safety relay is energized. This occurs through the series circuit of the normally closed flame detector switch, the warp switch heater, the equalizing resistor, the thermostat, the oven protectomatic and the motor throwout switch. The relay coil, when energized, will close the normally open relay contacts and the gas valve solenoid circuit is completed. Gas is allowed to flow to the burner to be ignited by the sparking electrodes.

After the burner is ignited, the sensor bar of the ignitor assembly expands and the flame detector switch is moved from its *normally closed* position to the *normally open* position. This opens the circuit to the warp switch heater as well as the relay coil. (The relay coil circuit is now completed through the closed relay points and the lockout resistor).

The dryer will continue in operation, responding to the control of the adjustable thermostat only, unless some of the following occurs:

If the gas control is opened by the drum protectomatic, the timer, the motor throw-out switch, the oven protectomatic, or the thermostat, the relay coil will be de-energize causing the relay points to open. This closes the solenoid valve. The valve will open only after the flame detector switch returns to its normally closed position and so re-energizes the relay coil and closes the relay contacts.

If the loading door is opened, the door switch will open the circuit to the timer motor, the drive motor and the gas solenoid (when the motor throw-out switch opens).

Should the burner not light the wrap switch heater will cause the warp switch contacts to open in about 30-75 seconds and the circuit to both the magnetic vibrator (sparking will stop) and the valve solenoid (shut off gas) will be opened, leaving the wrap switch heater to continue heating for the rest of the timed cycle (or until manually turned off). Before turning the timer back on, allow at least 5 minutes for the wrap switch to cool and return to its closed position.

Checking The McQuay-Norris "SIGI".
Automatic Ignition in the dryer takes place as the direct combination of many individual controls. Defects may be the result of the failure of one of these units, or maybe more. At least the important thing is to check the entire system so that effort and money is not wasted in the process of time — consuming parts changing. *Figure 144* shows a typical wiring diagram for a dryer using "SIGI" ignition.

Being familiar with the basic function of each component will insure the relationship pattern for quick and easy testing and diagnosis. Most of the various ignition systems can be checked out by "eye and ear" methods. The use of trouble shooting charts and the wiring diagram that is attached to each dryer is the surest way to check the system. Required to the job is a 115 volt test lamp, *Figure 145,* a neon tester, *Figure 146,* a 2½ volt test lamp and some insulated wire. Expose all controls and start out with this in mind: ALL TESTS ARE MADE WITH POWER AND GAS ON. It is easy to

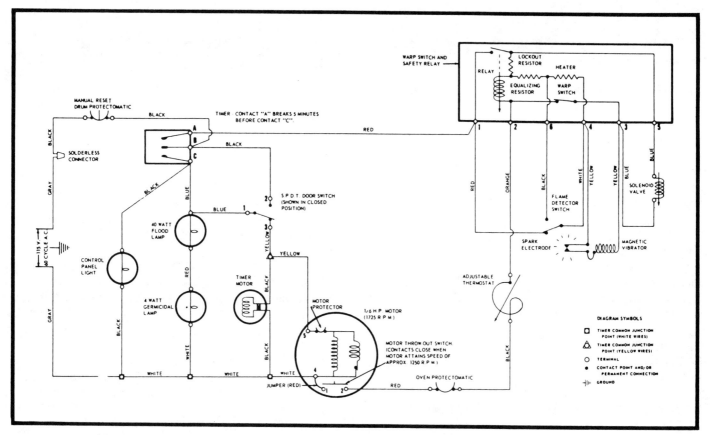

Figure 144

short out the components if care is not exercised. The tests that are to be made on the components in a dryer with a dryness control such as SEN-SITRON or SENTRY are made with TIME CYCLE control setting. Orderly checks should be made on the following system parts:

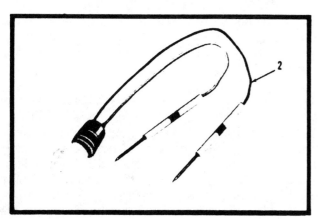

Figure 145

DRUM PROTECTOMATIC, with timer in OFF position, loading door open, connect the Neon test light between the B terminal (L1 on Sensitron) and the common terminal B (white wire) on the timer. NO LIGHT indicates that the cord or the drum protectomatic is "open". Reset the protectomatic and check the temperature. (Explained elsewhere).

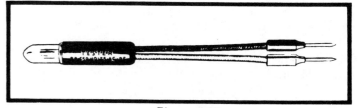

Figure 146

DOOR SWITCH CONTACTS 1 AND 2; The Neon light isplaced from C terminal (or M on the Sensi-tron) to the common terminal B (white wires) where a LIGHT indicates that the door switch is O.K. The lamps can be checked at this time, also.

TIMER: Turn timer to 5-minute position with loading door closed. Dryer motor should run and the drum revolving. If not: Place the Neon test light across terminals B and C on timer (Sensitron L1 and M). A light here indicates open points (contacts) Replace the control.

DOOR SWITCH CONTACTS 1 AND 3 AND THE MOTOR: The Neon light between the ommon terminal of white wires and the commmon terminal of yellow wires on the timer should LIGHT if the door switch is O.K. and the power is going to the motor. A test at 4 and 5 *on the motor* should not show a light between these terminals. A LIGHT indicates an open winding in the motor, replace.

To eliminate the checks for the next three items, make the following test. If they are serviceable, the neon test light across the terminals 1 and 2 of the power pack should light.

If no light is present, the timer, thermostat and motor throwout switch (with oven protectomatic) must be checked out individually, *before* proceding to the Flame Detector Switch check.

TIMER: If the Neon test light LIGHTS across the terminals A and B (H and L1-Sensitron) the contacts are open and the control must be replaced. Test is made with manual gas valve OFF' thermostat to HEAT and timer to the 30 minutes position.

THERMOSTAT: A LIGHT, when test light (Neon) is placed between (across) thermostat terminals, indicates an *open* thermostat; replace.

MOTOR THROW-OUT SWITCH AND OVEN PROTECTOMATIC: A test light between the black thermostat wire terminal to the common terminal of white wires on the timer will LIGHT if either the protectomatic or the throw-out switch is open. To further identify which, the test light is moved to the red wire on the protectomatic and the common terminal on the timer for the white wires. A LIGHT indicates the motor switch is open. NO LIGHT means the protectomatic is open.

FLAME DETECTOR SWITCH: FOR THIS TEST THE TIMER IS OFF. Remove orange lead from terminal 2, place a light (Neon) from the disconnected lead to the terminal 4. Now, turn the timer ON and if there is a LIGHT, the flame detector switch is serviceable. Turn timer off. Leave the orange off for the next test.

WARP SWITCH, HEATER, EQUALIZING RESISTOR AND THE RELAY COIL: Disconnect the yellow lead from terminal 4 of the power pack and then place the test light from disconnected orange wire (from last test) to terminal 2 of power pack.

Turn the timer on. NO LIGHT indicates a defective power pack. Replace. Turn timer off and reconnect wires.

WARP SWITCH CONTACTS: With the test light connected between the power pack terminals 1 and 3, turn timer on. A LIGHT will show warp switch O.K. but NO LIGHT means it is open. Turn the timer off and wait 5 minutes for the warp switch heater to cool. Retest and if still there is NO LIGHT, replace the power pack.

NOTE: Within 30-75 seconds the warp switch will open the circuit to magnetic vibrator and the gas valve; if this happens, turn off the timer and wait 5 minutes to repeat.

MAGNETIC VIBRATOR: TURN TIMER OFF. Take off the yellow wire from terminal 3. The Neon test light is placed between the disconnected lead and the terminal 2 of the power pack. Turn timer on. NO LIGHT condition indicates a bad magnetic vibrator.

Turn manual gas valve ON, Thermostat at HEAT, Timer to 30 minutes position for the next two tests.

SAFETY RELAY CONTACTS: A LIGHT between terminals 1 and 5 indicates an open condition. Replace.

GAS VALVE SOLENOID: Take off the blue lead from solenoid coil. Place a test light between this wire and the coil terminal. NO LIGHT means the solenoid coil is defective. Replace.

GAS BURNER AND OVEN SERVICE
By removing the top from the dryer, the oven area (including the burner) can be reached to make adjustments or replacements of the gas carrying components.

The combination gas valve and regulator assembly can be removed as shown in *Figure 147*. This unit directs the flow of the gas to the pilot and the main burner, when the two solenoids are energized (the "SIGI" has only one solenoid). The regulator section of the unit controls the incoming pressure of the gas, except when the gas used is LP. Then, the regulator is on the supply tank and the one the machine is blocked open.

To replace this unit be sure that the gas is shut off and the leads from the solenoid coils disconnected. On some manual controls it will be necessary to disconnect the 1/4" pilot tubing, as well.

By taking out the screws securing the assembly to the cross-angle, the unit may be lifted out when it is carefully turned to clear obstructions. Reassemble in reverse order.

Remove the screws that hold the oven to the wrapper sheet, slide it along the contour of the sheet and clear the slots in the wrapper. Then, turn the assembly so that the burner will clear the orifice fitting and lift it off. See *Figure 149.*

Figure 147

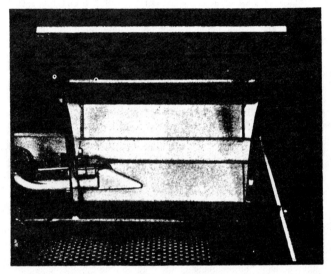

Figure 149

OVEN AND BURNER REMOVAL

To remove the oven, burner and pilot (ignitor), be sure to disconnect the leads from the oven protectomatic and the pilot switch. Also, the yellow leads from the Power Pac, if used, should be removed, at the Power Pac.

Some models have a 1/4" tubing from the shut-off to the pilot burner. Disconnect this on the side going to the burner.

After removing the screws holding main valve to the cross-angle, the oven bracket and the asbestos insulation can be removed. Two more metal screws hold the oven cover to the oven. Remove these and grasp the oven cover so that it can be pulled to the left, *Figure 148,* so as to disengage the oven cover angle from the wrapper sheet angle.

The burner assembly comes out after removing the screws holding it to the oven. If desired, the pilot or ignitor can be removed from the burner assemble by removal of two machine screws. Before reassembling, in reverse order, make sure that the main burner orified is put back in the burner. The "SIGI" burner assembly is shwon in *Figure 150.*

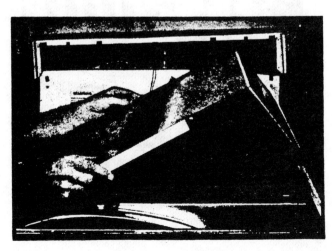

Figure 148

Figure 150

MAIN BURNER ORIFICE

To replace the orifice, shut off the gas, and disconnect the shut-off valve union nut. If it has a 1/4" pilot tubing to the filter, remove the compression nut. Take off the valve by removing the sheet metal screws that hold it to the cross-angle. Free it from the burner and then remove the orifice from the fitting. To reassemble, reverse the above instructions.

Pilot Orifice replacement can be accomplished by performing oven cover removal steps and then disconnect the shut-off tubing so that a wrench can be used to hold the orifice while taking the compression nut off with another wrench. At this point care should be exercised to avoid striking the ignitor sensor assembly. The orifice can now be removed by holding the pilot body and turning the orifice with another wrench. Reassemble in reverse order. Check for leaks and position of tubing. Replace the protectomatic leads.

CAST IRON BURNER AND OVEN REMOVAL

Prior models of the dryer in which the burner construction and the location of component parts are different than previously described, should be serviced in thefollowing manner.

Remove the pilot filter as shown in *Figure 151*. The gas pressure regulator can be taken off after taking out the screw holding it to the panel, *Figure 152*, but in some models the screw may be located on top of the rear cross bar flange. To free the solenoid valve and regulator, the valve union nut must be disconnected as shown in *Figure 153*.

The pilot orifice is removed through the access door by disconnecting the tubing, taking care that the pilot position is not disturbed. If this position is changed, the main burner may fail to light properly. *Figure 154* illustrates removal.

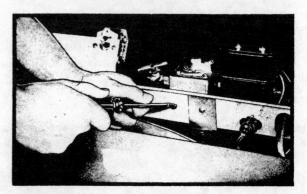

Figure 152

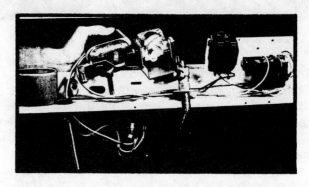

Figure 153

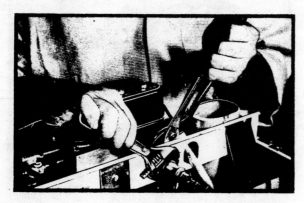

Figure 151

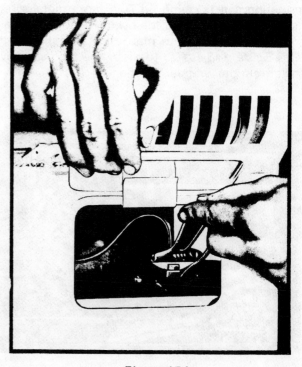

Figure 154

The main burner orifice removal requires the burner swing bail to be unlocked and the main burner tubing to be disconnected at the elbow connection. Pull out the tubing and remove the orifice. This replacement is installed in the reverse order. See *Figure 155.*

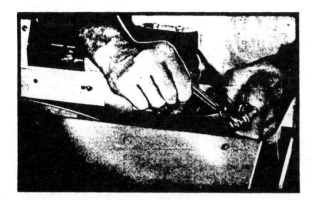

Figure 155

Figure 156 demonstrates the manner of removing wire leads from the oven protectomatic, while *Figure 157* shows the line between the pilot and the main burner being disconnected. The pilot switch wiring must also be removed and, in some cases the mercury bulb or the thermocouple should be removed from the burner (pilot). Take the screws out of the top oven cover and grasp the oven as illustrated in *Figures 158 and 159,* and lift off, being careful to slide it away from the tabs so that the burner will come with the cover, clearing the oven tabs from the slots on the wrapper sheet. The burner can be removed from the oven when the attaching clip is removed according to the model. Some locate it at the rear of the burner, while on others it is where the pilot burner fastens to the main burner.

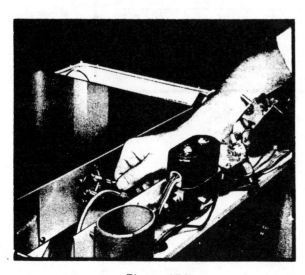

Figure 156

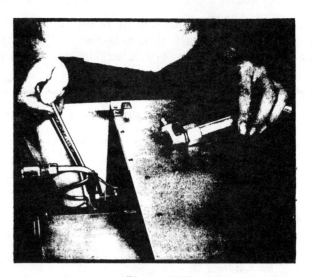

Figure 157

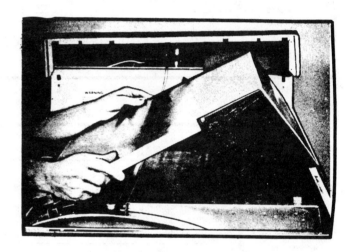

Figure 158

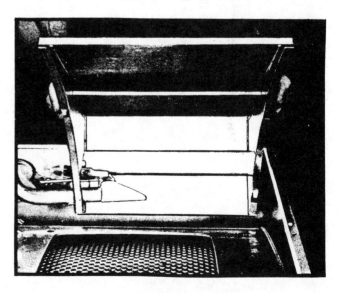

Figure 159

All units go back in position by reversing the foregoing steps. The chart, shown in *Figure 160,* lists the standards for the orifice size in late models as established at the factory. This is particularly important to follow if any changes in the gas supply are made. *Figure 161* is a chart for older models.

GAS AND ORIFICE SPECIFICATIONS FOR 18,000 B.T.U./HR. INPUT

The following table shows the standards for which orifices are adjusted when the dryers leave the factory:

	Natural	Mixed	Mfd.	Propane	Butane	LP Air	LP Air
B.T.U. Per Cu. Ft.	1075	800	530	2550	3200	525	1400
Specific Gravity	.65	.50	.60	1.54	2.00	1.16	1.42
Manifold Pressure	3½"	3½"	3½"	11"	11"	3½"	3½"
Main Burner Orifice	45	45	33	55	56	29	45
Pilot Orifice (McQuay-Norris controls)	*.022 .016	*.022 .016	*.031 .025	*.012 .010	*.012 .010	*.031 .025	*.022 .016
Pilot Orifice (Milw. Gas Spec. – Baso Controls)	.023	.023	.032	.012	.012	.032	.023

Gas and Orifice Specifications. Note: McQuay Norris orifice size numbers preceded by an * are for McQuay Norris Component ignition.

Figure 160

STANDARD BURNER SPECIFICATIONS FOR 18,000 BTU/HR INPUT

Type of Gas	BTU per Cu. Ft.	Pressure Inches	Specific Gravity	Pilot Orifice Number	No. of Ports in Pilot Orifice	Main Burner Orifice	Drill Size of Ports	No. of Ports in Main Burner
Manufactured	530	3½	.6	.028	2	33	36 DMS	38 Single Row
Mixed	800	3½	.6	.026	2	42	30 DMS	72 Double Row
Natural	1050	3½	.676	026	2	45	30 DMS	72 Double Row
Propane	2550	11	1.54	012	1	55	34 DMS	102 Double Row
Butane	3200	11	1.71	012	1	56	34 DMS	102 Double Row

Figure 161

DIAGNOSIS CHARTS

Probable Cause	Possible Remedy
*MANUAL IGNITION	
This section applies only to the ''Mercury-Vapor''' ignition system. When checking electrical components, the dryer should be plugged in and all controls should be in operation.	
*PILOT WILL NOT LIGHT	
Did not follow directions for lighting.	Follow directions on operator's card and/or on instructions attached to access panel.
Gas shut-off valves not open.	Open gas shut-off valves.
Pilot filter plugged.	Check pressure with manometer on inlet and outlet side of filter. Pressure drop of ½ inch is normal. Replace filter if necessary.
Pilot orifice plugged.	Remove orifice and visually check opening. Clean out, but do not ream hole.
*PILOT GOES OUT	
Low gas pressure.	Connect manometer at tap on valve. Pressure with main burner operating. should be 3½ inches water column pressure or 11 inches for L.P. gas. Adjust regulator to provide correct pressure.
Exhaust pipe off fan housing causing draft.	Visually check vent pipe and reposition.
Defective mercury element or latch assembly.	Remove latch mechanism cover and check engagement as covered in ''Mercury-Vapor'' *Page 69, Figure 127.*
Incorrect air circulation.	Improper air circulation caused by air leaks from improper fitting drum seals, improper fit of internal exhaust pipe to fan housing, fan housing cover leak, sticking vent hood flipper door, or clogged exhaust venting can result in back-drafts which affect pilot flame. If dryer is not vented, make sure exhaust restrictor was attached over interior exhaust pipe.
Air in gas line.	When condition occurred at time of installation, or immediately after, it is possible that problem resulted from air in the gas line.
*PILOT GOES OUT (CONT)	
Plugged orifice or pilot burner.	Check for dirty, or loose pilot orifice. At the same time also clean pilot burner primary air opening of any lint or foreign material. NOTE: When removing components, disconnect the pilot-to-filter tubing at orifice end. Then remove the one screw which secures the main burner mounting bracket to the valve bracket. The entire burner assembly can now be moved forward for access.
Bent or incorrectly positioned shields or brackets.	Visually check components and if found defective replace.

*Indicates condition applicable on only some models.

DIAGNOSIS CHARTS

Probable Cause	Possible Remedy
*PILOT LIGHT BURNING TOO HIGH OR TOO LOW	
Defective, or plugged orifice.	Pilot flame will burn in an abnormal manner. Replace orifice if defective or incorrect; clean if dirty. Never ream pilot orifice.
Incorrect gas pressure.	Check pressure with manometer. Pressure should be 3½ inches water column pressure on all gases except L.P. If incorrect, readjust regulator. Remove cover and turn slotted screw head counterclockwise to decrease pressure and clockwise to increase pressure. When L.P. units are involved, pressure should be 11 inches.
Pilot filter dirty.	Check pressure on filter. A drop of ½ inch through filter is normal. Clean or replace filter.
*PILOT BURNING WITH A RED OR YELLOW FLAME	
Lack of primary air.	Observe color of flame. Clean lint and dirt from primary air opening in burner body.
*PILOT LIT, DRYER IS RUNNING BUT MAIN BURNER DOES NOT LIGHT	
Plugged burner orifice.	Visually inspect orifice. Clean or replace if necessary.
Defective timer.	Check timer in the "on" position by placing a test light across terminals "A" and "B" or "H" and "L1". LIGHT indicates open contacts. Replace timer.
Thermostat contacts open.	With dryer turned on and thermostat in any position except "Air Fluff," place test light across terminals. LIGHT indicates contacts open. Replace defective thermostat.
Defective fan housing safety protectomatic, or heat chamber protectomatic.	Place neon test light across terminals of protectomatic. LIGHT indicates protectomatic defective. Replace.
Defective main solenoid coil or motor switch.	Place neon test light across terminals of solenoid coil. LIGHT indicates power to coil and motor throw-out switch contacts closed, coil is defective. NO LIGHT, check motor throw-out switch by checking across terminals 1 and 2 of motor. LIGHT indicates switch open. Replace or repair as necessary.
*AUTOMATIC IGNITION (SOLID STATE)	

This section applies only to the "Solid State" automatic ignition system.

When checking electrical components, the dryer should be plugged in and all controls should be in operation.

ALSO REFER TO SOLID STATE *PAGE 65.*

*Indicates condition applicable on only some models.

DIAGNOSIS CHARTS

Probable Cause	Possible Remedy
*MAGNETIC VIBRATER WILL NOT SPARK	
Place neon test light across Power Pac terminals 1 and 2. If light does NOT LIGHT, then check the following components.	
Defective timer.	Place a neon test light across terminals "A" and "B" or "H" and "L1." LIGHT indicates contacts open. Replace timer.
Defective thermostat.	Place a neon test light across terminals of control thermostat. LIGHT indicates contacts open. Replace thermostat.
Defective or "open" safety protectomatic.	Place neon test light across terminals of fan housing protectomatic. LIGHT indicates protectomatic open. Determine cause and if defective, replace.
Defective or open heat chamber protectomatic.	Place neon test light across terminals of heat chamber protectomatic. LIGHT indicates protectomatic open. Determine cause and if defective, replace.
Defective motor throw-out switch.	Turn timer off and disconnect brown lead from power pack. Turn timer on. Place neon test lamp from disconnected lead to terminal 1 of power pack. NO LIGHT indicates open motor throw-out switch. Remove motor and repair or replace switch. Reconnect brown lead to terminal 2.
Defective flame sensor switch.	Turn timer off, disconnect brown wire on terminal 1 of power pack. Turn timer on. Place Neon test lamp from disconnected lead to terminal 3 of power pack. LIGHT indicates sensor switch O.K. Turn timer off. Reconnect lead.
Defective or open warp switch.	Turn timer on and place neon test light across terminals 1 and 4 of power pack. LIGHT indicates warp switch contacts are open. Replace power pack.
Defective magnetic vibrator.	Since checks of flame sensor switch and warp switch above indicate you have electrical power at vibrator, failure for vibrator to work indicates component is defective. Check for "open" in lead wires or dirt particles on contact points. Replace magnetic vibrator if necessary. Turn timer off and wait 5 minutes for warp switch to cool before proceeding to next step. ***Note** Magnetic vibrator failure can be verified by attaching a 115V test service cord directly to vibrator ends. (Leads must be disconnected from valve and safety pack for test.)
*IGNITOR SPARKS BUT BURNER WILL NOT IGNITE	
Gas shut-off valve closed.	Visually check. Open valve.
Burner orifice plugged.	Check burner orifice. Clean lint and dirt from opening. Never ream orifice.
Defective power pack.	Check circuitry and replace as necessary.
Defective solenoid coil.	Turn dryer off. Place neon test light across blue wires on solenoid coil and restart dryer. LIGHT indicates power to coil. If coil will not energize to open valve, replace. This is a D.C. circuit. Do **NOT** test coil with 115V test cord.

*Indicates condition applicable on only some models.

SECTION 5

PARTS LISTS

The following parts lists are representative of the majority of the more popular parts used in servicing Hamilton Gas & Electric Dryers. Mainly they are shown as an aid in assembly sequence and to show the nomenclature of the various parts.

When ordering parts, always give the full model and serial number of the dryer. These numbers are found on a metal identification plate on the back of the machine or in the clothes door well.

GAS CARRYING & IGNITION PARTS FOR MODELS:
(3A8, 4A8)

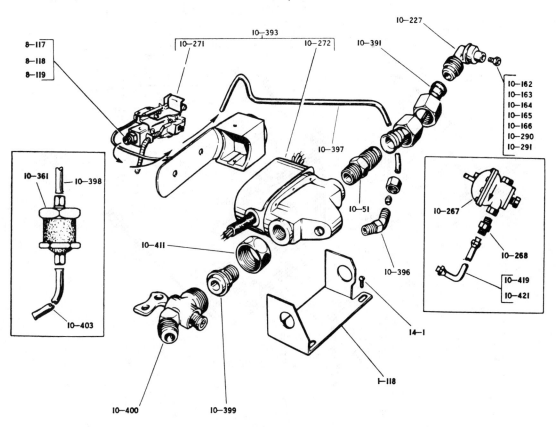

1-118	Control Support Bracket
8-117	Pilot Orifice Plug (mfg. gas)
8-118	Pilot Orifice Plug (natural, mixed, Propane & Butane)
8-119	Pilot Orifice Plug (LP gas)
10-51	Half Union Body
10-162	Orifice Plug (natural gas)
10-163	Orifice Plug, Blank (mfg., natural, or mixed)
10-164	Orifice Plug (Propane)
10-165	Orifice Plug (Butane)
10-166	Orifice Plug, Blank (L.P.)
10-227	Orifice Fitting
10-267	Gas Pressure Regulator
10-268	Union Body for Regulator
10-271	Switch, Safety Pilot & Lead Assembly
10-272	Twin Solenoid Valve Assembly less Part 10-271
10-282	Ignition Block Assembly (Not Shown)
10-290	Orifice Plug (mixed gas)
10-291	Orifice Plug (mfg. gas)
10-361	Gas Filter (mfg. gas)
10-391	½'' Tubing Assembly
10-393	Combination Dual Solenoid Valve
10-396	45° Brass Elbow
10-397	¼'' Tubing (all gas except mfg.)
10-398	¼'' Tubing, Filter to Pilot (mfg. gas)
10-399	Union Male Tail Piece
10-400	Shut-off Valve
10-403	¼'' Tubing, Solenoid to Filter (mfg. gas)
10-411	3/8'' Union Nut
10-419	½'' Tubing Assembly (LP gas)
10-421	½'' Tubing Assembly for Mfg., Natural & Mixed Gases
14-1	Attaching Screw (Procure Locally)

GAS CARRYING PARTS FOR MODELS (1000—G)

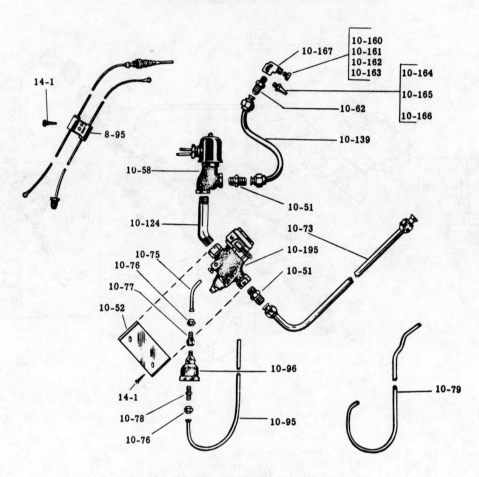

8-95	Junction Block and Couple Lead Assembly
10-51	Male Half Union Body 1/2" S.A.E. x 3/8" I.P.T.
10-52	Bracket for Baso Valve
10-58	Magnetic Valve
10-62	Male Half Union Body 1/2" S.A.E. x1/4" I.P.T.
10-73	1/2" O.D. Aluminum Tubing Assembly
10-75	1/4" O.D. Aluminum Tubing, 4" long, Plus Sleeve
10-76	1/4" S.A.E. Special Short Type Union Shell
10-77	Female Half Union Body, Male 1/4" S.A.E. x Female 1/8" I.P.T.
10-78	Male Half Union Body, 1/4" S.A.E. x 1/8" I.P.T.
10-79	1/4" O.D. Aluminum Tubing Plus Sleeves for Natural Mixed and L.P. Gas
10-95	1/4" O.D. Aluminum Tubing Plus Sleeve 26-1/2" long, for Mfgd. Gas
10-96	Filter for Pilot Burner
10-124	Nipple
10-139	Aluminum Tubing Assembly
10-160	Orfice Plug, Mfgd. Gas
10-161	Orfice Plug, Natural Gas
10-162	Orfice Plug, Mixed Gas
10-163	Orfice Plug, Blank
10-164	Orfice Plug, Propane Gas
10-165	Orfice Plug, Butane Gas
10-166	Blank Plug, L.P. Gas
10-167	Orfice Fitting
10-195	Base Valve with Handle Extension

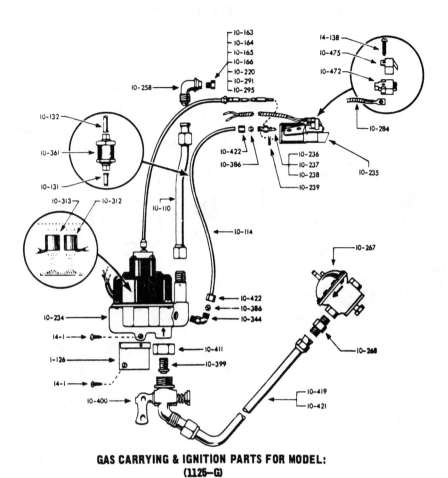

GAS CARRYING & IGNITION PARTS FOR MODEL:
(1125–G)

1-126	Support Bracket	10-268	Union Body for Regulator
10-110	½'' Tubing	10-284	Sheathed Cable
10-114	¼'' Tubing	10-291	Orifice Plug (mfg. gas)
10-131	¼'' Tubing (Mfg. Gas)	10-295	Orifice Plug (natural & mixed)
10-132	¼'' Tubing (Mfg. Gas)	10-312	Replacement Solenoid Coil (Pilot)
10-163	Orifice Plug Blank (natural, mixed & mfg.)	10-313	Replacement Solenoid Coil (Main)
10-164	Orifice Plug (Propane)	10-344	Elbow
10-165	Orifice Plug (Butane)	10-361	Pilot Filter (mfg. gas)
10-166	Orifice Plug Blank (LP)	10-386	Compression Fitting
10-220	Velocity Reducing Spud	10-399	Union Male Tail Piece
10-234	Dual Solenoid Valve & Switch Assembly	10-400	Shut-off Valve
10-235	Pilot Light Assembly, Shield, Cable, & Coil & Block Assembly	10-411	3/8'' Union Nut
		10-419	½'' Tubing (LP gas)
10-236	Pilot Orifice (natural & mixed)	10-421	½'' Tubing (except LP gas)
10-237	Pilot Orifice (Mfg.)	10-422	Nut for Compression Fitting
10-238	Pilot Orifice (LP)	10-472	Ignitor Coil
10-239	Pilot Orifice Clip	10-475	Shield for Ignitor Coil
10-258	Orifice Fitting	14-1	Screw
10-267	Gas Pressure Regulator	14-138	Screw for Attaching Ignitor Coil & Block Assembly

GAS CARRYING PARTS FOR MODELS (250—G, 152G)

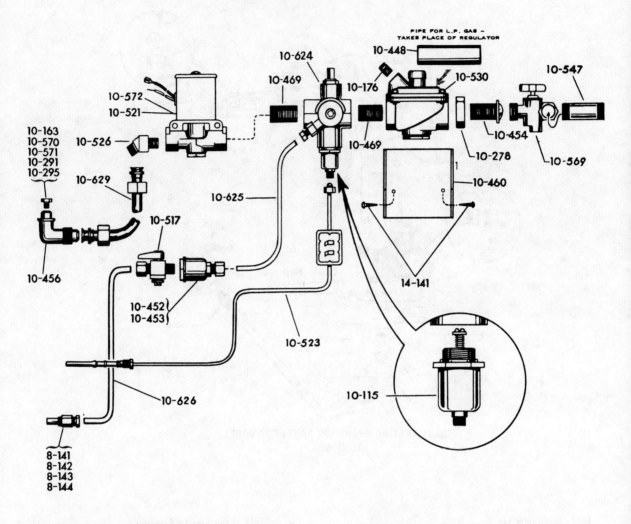

8-141	Pilot Orifice Plug, Mfd. Gas (.032)
8-142	Pilot Orifice Plug, L.P. Air Gas at 525 B. T. U. (.032)
8-143	Pilot Orifice Plug, Nat., Mixed and L.P. Air at 1400 B.T.U. (.032)
8-144	Pilot Orifice Plug, L.P. (.012)
10-115	Magnetic Hood for Baso Valve
10-163	Main Burner Orifice Blank for all Gases
10-176	Vent Limiting Device
10-224	Coil for 10-521 General Controls Solenoid
10-278	Union Nut
10-291	Main Burner Orifice Plug, Mfd., Gas (#33 Drill)
10-295	Main Burner Orifice Plug, Nat., and Mixed Gas (#44 Drill)
10-386	Compression Fitting for 1/4" O.D. Tubing (Replacement)
10-422	Nut for Fitting #10-386 (Replacement)
10-448	Coupling for L.P. Gas Only
10-452	Gas Filter for Mfd. Gas Only
10-453	Gas Filter for All Gases Except Mfd. Gas
10-454	Union Nipple for Shut-Off Valve
10-456	Orifice Fitting
10-460	Mounting Bracket for Gas Controls
10-469	3/8" Nipple
10-517	Pilot Shut-Off Valve
10-521	Solenoid Valve (Alternate with #10-572)
10-523	Junction Block and Thermo Lead Assembly
10-526	3/8" N.P.T. x 1/2" S.A.E. 45 Degree Elbow
10-530	Gas Pressure Regulator
10-541	Magnetic Hood for Baso Valve (Replacement)
10-547	3/8" x 1-1/2" Black Nipple - Gas Inlet
10-569	Shut- Off Valve
10-570	Main Burner Orifice for Propane (#55 Drill)
10-571	Main Burner Orifice for Butane (#56 Drill)
10-572	Solenoid Valve (Alternate with #10-521)
10-624	Baso Valve
10-522	Baso Valve (Model 250G)
10-625	1/4" O.D. Tubing, Baso to Gas Filter
10-626	1/4" O.D. Tubing, Pilot Valve to Burner
10-629	1/2" O.D. Tubing, Solenoid to Burner
10-632	Coil for #10-572 McQuay Norris Solenoid
14-141	Screw

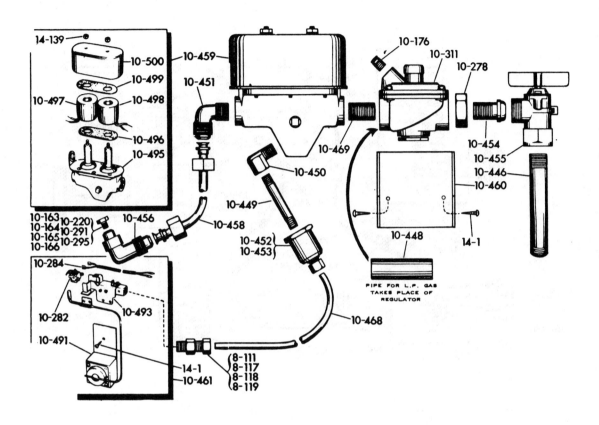

GAS CARRYING & IGNITION PARTS FOR MODELS:
(260G, 360G)

8-111	Pilot Orifice Plug, Propane Gas (.0145)	10-451	3/8'' H.P. T x ½'' S.A.E. 90° Elbow
8-117	Pilot Orifice Plug, Mfd. Gas (.028)	10-452	Gas Filter for Mfd. Gas Only
8-118	Pilot Orifice Plug, Natural & Mixed Gas, Propane Air & Butane Air Gas (.026)	10-453	Gas Filter for All Gases except Mfd. Gas
		10-454	Union Nipple for Shut Off Valve
8-119	Pilot Orifice Plug, Butane Gas (.012)	10-455	Shut Off Valve
10-163	Main Burner Orifice Blank Plug for Mfd., Natural & Mixed Gas and Propane Air	10-456	Main Burner Orifice Plug Fitting
		10-458	Tubing Assembly from Solenoid Valve to Burner
10-164	Main Burner Orifice Plug, Propane Gas (#55 Drill)	10-459	White-Rodgers Dual Solenoid Valve Assembly
10-165	Main Burner Orifice Plug, Butane Gas (#56 Drill)	10-460	Mounting Bracket for Gas Controls
10-166	Main Burner Orifice Blank Plug, Propane & Butane Gas	10-461	White-Rodgers Safety Switch and Pilot Assembly
10-176	Vent Limiting Device for Regulator	10-468	¼'' O.D. Aluminum Tubing from Filter to Pilot Burner
10-220	Main Burner Orifice Blank Plug, Butane Air	10-469	Nipple from Regulator to Dual Valve
10-278	Union Nut for Shut Off Valve	10-491	White-Rodgers Pilot Device Assembly
10-282	White-Rodgers Ignitor Coil Assembly	10-493	White-Rodgers Pilot Burner Block with Shield
10-284	White-Rodgers Sheathed Lead	10-495	White-Rodgers Cast Base for Dual Valve with Studs for Coils
10-291	Main Burner Orifice Plug, Mfd. Gas (#33 Drill)		
10-295	Main Burner Orifice Plug, Natural & Mixed Gas	10-496	White-Rodgers Lower Fibre Washer
10-311	Gas Pressure Regulator (All Gases except LP)	10-497	White-Rodgers Main Solenoid Valve Coil
10-386	Compression Fitting for ¼'' O.D. Tubing (Not Illustrated)	10-498	White-Rodgers Pilot Solenoid Valve Coil
		10-499	White-Rodgers Upper Fibre Washer
10-422	Nut for Compression Fittings #10-386 (Not Illustrated)	10-500	White-Rodgers Cover for Dual Valve Assembly
10-446	3/8'' Bent Pipe	14-1	Attaching Screw
10-448	Coupling for LP Gas Only	14-139	White-Rodgers Dual Valve Cover Attaching Nut
10-449	1/8'' x 2'' Male Nipple Extension for Gas Filter		
10-450	1/8'' I.P. Street Elbow		

GAS CARRYING & IGNITION PARTS FOR MODELS:
(260G, 360G)

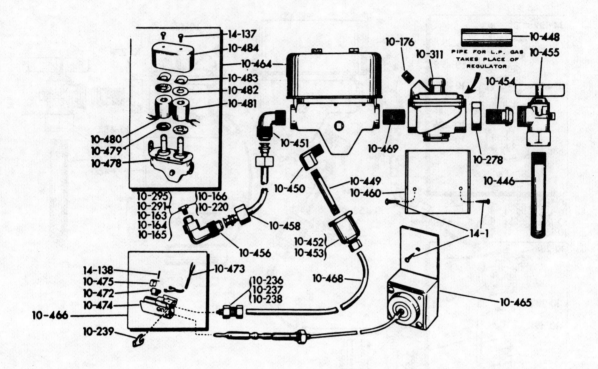

10-163	Main Burner Orifice Plug, Mfd., Natural & Mixed Gas & Propane Air	10-456	Main Burner Orifice Fitting
10-164	Main Burner Orifice Plug, Propane Gas (#55 Drill)	10-458	Tubing Assembly from Solenoid Valve to Burner
10-165	Main Burner Orifice Plug, Butane Gas (#56 Drill)	10-460	Mounting Bracket for Gas Controls
10-166	Main Burner Orifice Blank Plug, Propane & Butane Gas	10-464	Perfex Dual Valve Assembly
10-176	Vent Plug	10-465	Perfex Pilot Device Assembly
10-220	Main Burner Orifice Blank Plug, Butane Air	10-466	Perfex Pilot Burner Assembly with Ignitor, Flame Shield, Mounting Bracket & Sheathed Lead
10-236	Perfex Pilot Orifice, Natural & Mixed Gas (.020)	10-468	¼'' O.D. Aluminum Tubing from Filter to Pilot Burner
10-237	Perfex Pilot Orifice, Mfd. Gas (.026)	10-469	3/8'' Close Nipple from Regulator to Dual Valve
10-238	Perfex Pilot Orifice, Propane & Butane Gas (.0145)	10-472	Perfex Ignitor Coil Assembly
10-239	Perfex Orifice Clip for Use with Perfex Pilot Orifice	10-473	Perfex Sheathed Lead
10-278	Union Nut for Shut Off Valve	10-474	Perfex Pilot Burner Assembly
10-291	Main Burner Orifice Plug, Mfd. Gas (#33 Drill)	10-475	Perfex Ignitor Shield
10-295	Main Burner Orifice Plug, Natural & Mixed Gas (#44 Drill)	10-478	Perfex Cast Base for Dual Valve with Studs for Coils
10-311	Pressure Regulator (All Gases except LP)	10-479	Perfex Lower Fibre Washers for Coils on Dual Valve Assembly
10-386	Compression Fitting for ¼'' O.D. Tubing (Not Illustrated)	10-480	Perfex Main Solenoid Valve Coil
10-422	Nut for Compression Fitting #10-386 (Not Illustrated)	10-481	Perfex Pilot Solenoid Valve Coil
10-446	3/8'' Bent Pipe	10-482	Perfex Upper Fibre Washer for Coils on Dual Valve Assembly
10-448	Coupling for LP Gas Only	10-483	Perfex Solenoid Valve Holding Clips
10-449	Male Nipple Extension for Gas Filter, 1/8'' I.P. x 2''	10-484	Perfex Dual Valve Cover
10-450	1/8'' I.P. Street Elbow	14-1	Attaching Screw
10-451	3/8'' N.P.T. x ½'' S.A.E. 90° Elbow	14-137	Perfex Dual Valve Cover Attaching Screw
10-452	Gas Filter for Mfd. Gas Only	14-138	Perfex Ignition Coil Assembly Attaching Screw
10-453	Gas Filter for All Gases except Mfd. Gas		
10-454	Union Nipple for Shut Off Valve		
10-455	Shut Off Valve		

GAS CARRYING & IGNITION PARTS FOR MODELS:
(2M1, 3M1, 93M1, 4M1)

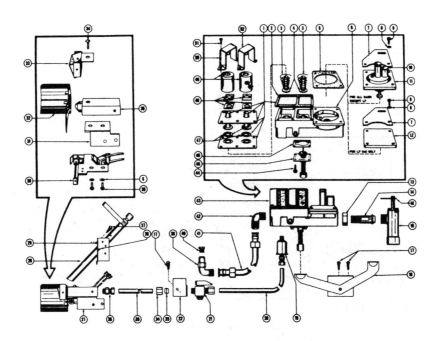

1. 10-966   Kit, Valve Plate
2. 10-967   Gasket, Valve Body
3. 10-968   Spring, Plunger Return
4. 10-969   Plunger, Solenoid
5. 10-970   Gasket, Regulator
6. 10-979   Body, Solenoid & Regulator
7. 10-954   Bracket, Mounting
8. 14-158   Lockwasher
9. 14-212   Screw, Regulator Cover
10. 10-972   Device, Vent Limiting
11. 10-973   Kit, Regulator (for all gases except L.P.)
12. 10-930   Cover, Regulator (for L.P. gas only)
13. 10-278   Nut, Union
14. 10-925   Nipple, Union
15. 10-846   Handle, Shut-off Valve
16. 10-843   Valve, Shut-off
17. 14-100   Screw, 3/8" #10 Sheet Metal
18. 10-856   Bracket & Lever, Plunger Activator
19. 10-848   Filter, Pilot (all gases)
20. 10-858   Tubing, ¼" O.D.
21. 10-871   Valve, Pilot Shut-off
22. 10-987   Bracket, Pilot Tubing Support
23. 10-386   Gland, Compression Fitting (for ¼" O.D. tubing)
24. 10-422   Nut, Compression Fitting (for ¼" O.D. tubing)
25. 10-859   Tubing, ¼" O.D.
26. 10-934   Orifice (.016), Natural & Mixed Gas Pilot
         10-933   Orifice (.025), Mfg. & L.P. Air (at twt BTU) Gas Pilot
         10-935   Orifice (.010), L.P. Gas Pilot
27. 10-936   Ignitor & Burner, Pilot
28. 8-179   Tube, Lighting

29. 8-178   Bracket, Lighting Tube
30. 10-937   Switch & Lead Assembly, Pilot
31. 10-977   Bracket, Pilot Mounting
32. 10-983   Shield, Pilot
33. 10-982   Shield, Pilot Flame
34. 14-216   Screw, #540 x ½" Machine
35. 10-978   Body, Pilot
36. 14-157   Screw, Switch to Pilot
37. 14-171   Screw, ½" #8 Sheet Metal
38. 14-214   Key, Cotter
39. 10-456   Fitting, Orifice
40. 10-162   Orifice (#45 Drill .082), Natural & Mixed Gas Main
         10-163   Orifice, Main Burner Blank
         10-291   Orifice (#33 Drill .113), Mfg. Gas Main
         10-570   Orifice (#55 Drill .052), Propane Gas Main
         10-571   Orifice (#56 Drill .0465), Butane Gas Main
41. 10-515   Tubing, ½" O.D.
42. 10-451   Elbow, 3/8" N.P.T. x ½" S.A.E. 90°
43. 10-928   Valve & Regulator, Gas (for all gases except L.P.)
         10-929   Valve & By-passed Regulator, Gas (for L.P. gas only)
44. 14-213   Screw, Valve Plate
45. 10-580   Plunger Assembly, Gas Valve Operating
46. 10-581   Gasket, Plunger
47. 10-974   Gasket, Valve Plate
48. 10-976   Spring, Solenoid Coil
49. 10-958   Coil, Solenoid Pilot & Main
50. 10-975   Bracket, Main Solenoid Hold-down
51. 14-213   Screw, Valve Plate to Body
52. 10-1105   Bracket, Pilot Solenoid Hold-down

GAS CARRYING & IGNITION PARTS FOR MODELS:
(3A8, 4A8)

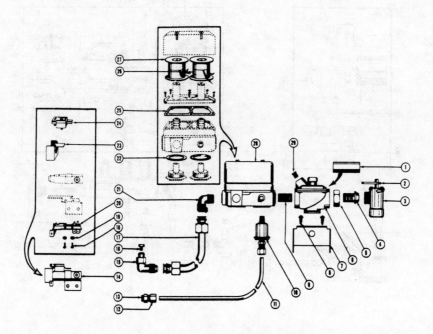

1.	10-448	Coupling (for LP gas)
2.	10-846	Handle, Shut-off Valve
3.	10-843	Valve, Shut-off
4.	10-454	Nipple, Union
5.	10-278	Nut, 3/8'' Union
6.	10-530	Regulator, General Controls, Gas Pressure
	10-695	Regulator, Thermac, Gas Pressure
7.	10-844	Bracket, Control Mounting
8.	14-141	Screw, ½'' #10 Sheet Metal
9.	10-469	Nipple, 3/8'' Close
10.	10-847	Filter, Mfg. Gas Pilot
	10-848	Filter, Natural, Mixed & LP Pilot
11.	10-849	Tubing, ¼'' O.D.
12.	10-422	Nut, Compression Fittings (for ¼'' O.D. Tubing)
	10-386	Gland, Compression Fitting (for ¼'' O.D. Tubing)
13.	10-574	Orifice (.020), McQuay Norris Component Pilot (for natural, mixed, & LP air at 1400 BTU)
	10-575	Orifice (.031), McQuay Norris Component Pilot (for mfg. gas & LP air at 500 BTU)
	10-576	Orifice (.012), McQuay Norris Component Pilot (for LP low and high altitude)
14.	10-562	Ignitor & Burner, McQuay Norris Component Pilot

15.	10-456	Fitting, Main Burner Orifice
16.	10-162	Orifice, (#45 Drill .082) Main Burner (for natural and mixed gas)
	10-163	Orifice, Main Burner Blank (for mfg., natural & mixed gas & Propane Air)
	10-291	Orifice, (#33 Drill .113) Main Burner (for mfg. gas)
	10-570	Orifice, (#55 Drill .052) Main Burner (for Propane)
	10-571	Orifice, (#56 Drill .0465) Main Burner (for Butane)
17.	10-515	Tubing, ½'' O.D., Solenoid Valve to Main Burner
18.	14-157	Screw, Switch to Pilot
19.	14-158	Washer, #6 Spring
20.	10-607	Switch, McQuay Norris Component Ignitor
21.	10-451	Elbow, 3/8'' N.P.T. x ½'' S.A.E. 90°
22.	10-590	Gasket, McQuay Norris Component Solenoid Plug
23.	10-605	Shield, McQuay Norris Component Pilot
24.	10-585	Coil & Block, McQuay Norris Component Ignitor
25.	10-579	Gasket, McQuay Norris Component Upper Solenoid
26.	10-729	Coil, McQuay Norris Component Pilot
27.	10-863	Coil, McQuay Norris Component, Main
28.	10-845	Valve, McQuay Norris Component Dual
29.	10-176	Vent Limiting Device (for General Controls Regulator only)

GAS CARRYING & IGNITION PARTS FOR MODELS:
(3A1, 4A1, 3A9, 4A9)

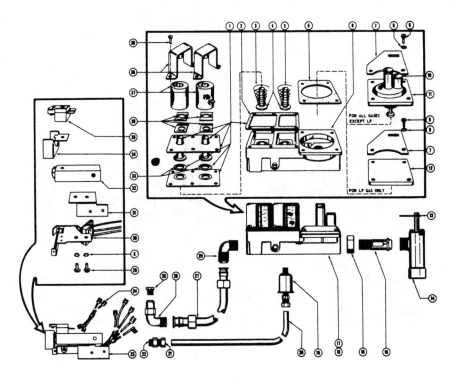

1.	10-966	Kit, Valve Plate
2.	10-967	Gasket, Valve Body
3.	10-968	Spring, Plunger Return
4.	10-969	Plunger, Solenoid
5.	10-970	Gasket, Regulator
6.	10-971	Body, Solenoid & Regulator
7.	10-954	Bracket, Mounting
8.	14-158	Washer, Lock
9.	14-212	Screw, Regulator Cover
10.	10-972	Device, Vent Limiting
11.	10-973	Kit, Regulator (for all gases except LP)
12.	10-930	Cover, Regulator (for LP gas only)
13.	10-846	Handle, Shut-off Valve
14.	10-843	Valve, Shut-off
15.	10-925	Nipple, Union
16.	10-278	Nut, Union
17.	10-926	Valve & Regulator Assembly (for all gases except LP)
18.	10-927	Valve & By-passed Regulator Assembly (for LP gas only)
19.	10-848	Filter, Pilot (all gases)
20.	10-849	Tubing, ¼" O.D.
21.	10-422	Nut, Compression Fitting
	10-386	Gland, Compression Fitting

22.	10-934	Orifice (.016), Pilot (for natural & mixed gas)
	10-933	Orifice (.025), Pilot (for mfg. & L.P. air at 525 BTU)
	10-935	Orifice (.010), Pilot (for LP. gas)
23.	10-931	Ignitor & Burner, Pilot
24.	10-955	Wiring, Ignitor Coil
25.	10-162	Orifice (No. 45 drill .082), Main Burner (for natural and mixed gas)
	10-163	Orifice, Main Burner Blank (for all gases)
	10-291	Orifice (No. 33 drill .113), Main Burner (for mfg. gas)
	10-570	Orifice (No.55 drill .052), Main Burner (for Propane gas)
	10-571	Orifice (No. 56 drill .0465), Main Burner (for Butane)
26.	10-456	Fitting, Main Burner Orifice
27.	10-515	Tubing, ½" O.D. Solenoid Valve to Main Burner
28.	10-451	Elbow, 3/8" N.P.T. x ½" S.A.E. 90°
29.	14-157	Screw, Switch to Pilot
30.	10-932	Switch & Lead Assembly, Pilot
31.	10-977	Bracket, Pilot Mounting
32.	10-978	Body, Pilot
33.	10-974	Gasket, Valve Plate
34.	10-605	Shield, Pilot
35.	10-585	Coil & Block, Ignitor
36.	10-976	Spring, Solenoid Coil
37.	10-958	Coil, Solenoid Pilot & Main
38.	10-975	Bracket, Solenoid Hold-down
39.	14-213	Screw, Valve Plate to Body

REDI–MITE–CONSTANT PILOT ASSEMBLY
MODELS: DM205W1 DM305W1, DM405W1,
DM495W1, 3M3, 4M3, 94M3

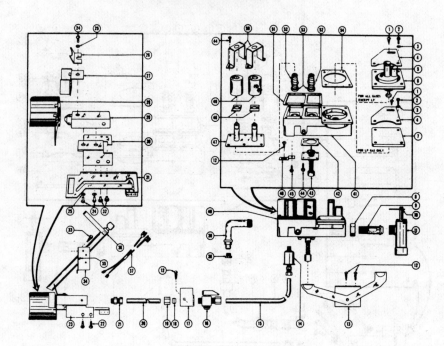

1.	14-134	Screw, No. 8 x 3/8" Valve Attaching
2.	14-212	Screw, Regulator Cover
3.	14-158	Washer, Lock
4.	10-954	Bracket, Mounting
5.	10-972	Device, Vent Limiting
6.	10-973	Kit, Regulator (For all gases except L.P.)
7.	10-930	Cover, Regulator (For L.P. gas)
8.	10-278	Nut, Union
9.	10-925	Nipple, Union
10.	10-846	Handle, Shut-off Valve
11.	10-843	Valve, Shut-off
12.	14-100	Screw, No. 10 x 3/8"
13.	10-856	Bracket and Lever, Plunger Activator
14.	10-848	Filter, Pilot (all gases) (without fittings)
15.	10-1175	Tubing, 1/4" O.D.
16.	10-871	Valve, Pilot Shut-off (with fittings)
17.	10-1178	Bracket, Pilot Tubing Support
18.	10-386	Gland, Compression Fitting (For 1/4" O.D. tubing)
19.	10-422	Nut, Compression Fitting (For 1/4" O.D. tubing)
20.	10-1176	Tubing, 1/4" O.D.
21.	10-934	Orifice (.016), Nat. and Mixed Gas Pilot
	10-933	Orifice (.025), Mfg. and L.P. Air (at 525 BTU) Gas Pilot
	10-935	Orifice (.010), L.P. Gas Pilot
22.	14-217	Screw, 8-32 x 3/8" Pilot Attaching
24.	14-216	Screw, 5-40 x 1/4"
25.	14-325	Washer, Lock
26.	10-1128	Bracket, Lighting Tube
27.	10-605	Shield, Pilot Flame
28.	10-983	Shield, Pilot

29.	10-978	Body, Pilot
30.	10-1321	Bracket, Pilot Mounting
31.	10-1323	Lead Assembly, Pilot Switch
32.	10-1322	Switch and Flame Sensor Assembly, Pilot
33.	14-326	Screw, 6-32 x 3/16" and Lockwasher
34.	14-171	Screw, No. 8 x 1/2" Sheet Metal
35.	8-178	Bracket, Lighting Tube
36.	14-214	Key, Cotter (1/16" x 1")
37.	8-197	Tube Assembly, Lighting
38.	10-162	Orifice (No. 45 Drill .082) Natural and Mixed Gas, Main
	10-163	Orifice, Main Burner, Blank
	10-291	Orifice (No. 33 Drill .113) Mfg. Gas, Main
	10-570	Orifice (No. 55 Drill .052)
	10-571	Orifice (No. 56 Drill .0465) Butane Gas, Main
39.	10-1383	Fitting, Orifice
41.	10-1324	Body, Solenoid & Regulator
42.	10-581	Gasket, Plunger
43.	10-580	Plunger Assembly, Gas Valve
44.	14-213	Screw, Valve Plate
45.	14-204	Screw, 10-32 x 3/8"
46.	10-1198	Bracket, Valve Support
47.	10-1150	Kit, Valve Plate
48.	10-976	Spring, Solenoid Coil
49.	10-1197	Coil, Solenoid Pilot & Main
50.	10-1149	Bracket, Solenoid Hold-down
51-	10-967	Gasket, Valve Body
52.	10-1162	Spring, Plunger Return
53.	10-1165	Plunger, Solenoid
54.	10-970	Gasket, Regulator

DIRECT INSTANT (SIGI) IGNITION CONTROL ASSEMBLY
MODELS: 4S3, 94S3, 5S3, DS305W1, DS405W1,
DS495W1, DS505W1, DS505C1

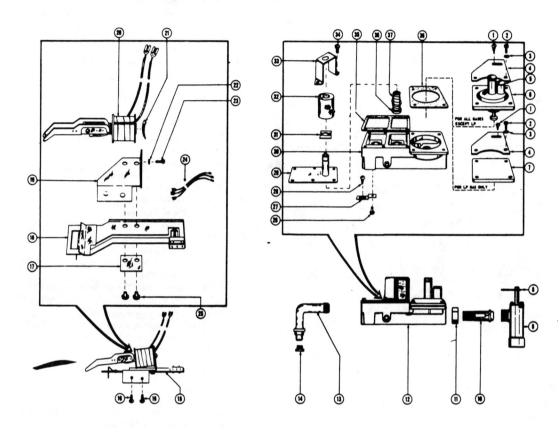

1.	14-134	Screw, #8 - 3/8" Valve Attaching
2.	14-212	Screw, Regulator Cover
3.	14-158	Washer, Lock
4.	10-954	Bracket, Mounting
5.	10-972	Device, Vent Limiting
6.	10-973	Kit, Regulator (for all gases except L.P.)
7.	10-930	Cover, Regulator (for L.P. gas)
8.	10-846	Handle, Shut-off Valve
9.	10-843	Valve, Shut-off
10.	10-925	Nipple, Union
11.	10-278	Nut, Union
13.	10-1383	Fitting, Orifice
	10-1180	Fitting, Orifice–4S3, 94S3, 5S3
14.	10-162	Orifice, Burner, Natural & Mixed Gas (#45 Drill .082)
	10-163	Burner Orifice, All Gases, Blank
	10-291	Burner Orifice, Mfg. Gas (#33 Drill .113)
	10-570	Burner Orifice, Propane Gas (#55 Drill .052)
	10-571	Burner Orifice, Butane Gas (#56 Drill .0465)
16.	14-217	Screw, 8-32 x 3/8" Ignitor Attaching
17.	10-1329	Plate, Sensor
18.	10-1328	Switch & Sensor Assembly, Ignitor

19.	10-1327	Bracket, Ignitor Mounting
20.	10-1325	Coil & Electrode Assembly, Ignitor (for all gases except L.P.)
	10-1427	Coil & Electrode Assembly, Ignitor (for L.P. gas)
21.	10-1326	Spring, Solenoid Coil
22.	14-328	Washer, Lock, #6
23.	14-327	Screw, Machine 6-32 x 3/8"
24.	10-1330	Lead Assembly, Ignitor
25.	14-326	Screw & Lockwasher
26.	14-204	Screw, 10-32 x 3/8"
27.	10-1198	Bracket, Valve Support
28.	14-100	Screw, #10 x 3/8"
29.	10-1252	Kit, Valve Plate
30.	10-1253	Body, Solenoid & Regulator
31.	10-976	Spring, Solenoid Coil
32.	10-1197	Coil, Solenoid
33.	10-1149	Bracket, Solenoid Hold-down
34.	14-213	Screw, Valve Plate to Body
35.	10-967	Gasket, Valve Body
36.	10-1165	Plunger, Solenoid
37.	10-1162	Spring, Plunger Return
38.	10-970	Gasket, Regulator

DRUM & DRIVE PARTS FOR MODELS:
(DE205W1, DE305W1, DE405W1, DE505W1, DE505C1)

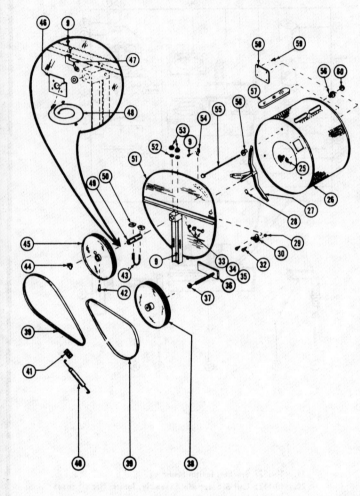

9.	14-100	Screw, #10 x 3/8''
25.	14-379	Nut, Tinnerman
26.	6-57	Drum Assembly, with Cleanout Doors, Tie Bolts, Nuts & Baffles—DE205W1, DE305W1, DE405W1
	6-91	Drum Assembly, with Cleanout Doors, Tie Bolts, Insulators, Nuts & Baffles—DE505W1, DE505C1
27.	6-10	Spider, Drum—DE205W1, DE305W1, DE405W1
	6-74	Spider, Drum—DE505W1, DE505C1
28.	14-373	Clip, Tinnerman—DE505W1, DE505C1
29.	14-225	Washer, Steel
30.	10-1233	Protectomatic, Drum
32.	14-134	Screw, #8 x 3/8''
33.	14-347	Lockwasher
34.	14-346	Locknut, ¼'' - 20
35.	14-345	Adjusting Screw, ¼'' - 20 x ¾''
36.	12-11	Idler Bar
37.	12-10	Ring, Truarc
38.	11-93	Pulley, Intermediate
39.	11-14	Drive Belt
40.	12-4	Tension Spring
41.	12-5	Plate, Keyhole
42.	14-22	Pulley Setscrew, 3/8'' - 16 x 5/8''
43.	5-19	U-Bolt, Bearing
44.	14-244	Wire Clamp—DE505W1, DE505C1
45.	11-51	Drum Pulley, Without Setscrew—DE205W1, DE305W1, DE505W1
	11-98	Drum Pulley, Without Setscrew—DE505W1, DE505C1
46.	10-1307	Slip Ring Assembly—DE505W1, DE505C1
47.	10-1302	Contact Assembly—DE505W1, DE505C1
48.	10-1485	Drum Sensing Harness—DE505W1, DE505C1
49.	5-168	Main Bearing
50.	5-172	Adjusting Block
51.	5-163	Head, Rear Drum Case
52.	14-61	Washers, U-Bolt
53.	14-60	Hex Nut, 5/16'' - 18
54.	8-11	Clip, Heater Fastening
55.	6-7	Bolt, Drum Tie—DE205W1, DE305W1, DE405W1
	6-76	Bolt, Drum Tie—DE505W1, DE505C1
56.	6-75	Bushing, Insulator—DE505W1, DE505C1
57.	6-28	Cover, Drum Cleanout
58.	6-56	Cover, Rear Drum Cleanout
59.	14-128	Screw, #10 x ½''
60.	14-90	Hex Nut, 5/16'' - 18
62.	14-20	Hex Nut, #10 - 32 (Attach Heater Leads)
63.	8-13	Screen, Heater
64.	8-1	Heater Assembly (With Screen)
65.	8-21	Reflector, Heater

DRUM ASSEMBLY FOR MODELS:
(3E3, 94E3, 5E3)

26.	6-25	Drum Assembly (With Cleanout Door)—3E3, 94E3
	6-48	Drum Assembly (With Cleanout Door but without Sensing Plates)—5E3
50.	6-7	Bolt, Drum Tie
51.	6-10	Spider, Drum—3E3, 94E3
	6-41	Spider, Drum—5E3
52.	10-1319	Sleeve, Insulated—5E3
53.	10-1320	Connector, Wire—5E3
54.	14-205	Screw, #10 x ¼''—5E3
55.	6-46	Bushing, Insulator—5E3
56.	6-43	Plate, Sensing—5E3
57.	6-45	Insulator, Sensing Plate—5E3
58.	6-28	Cover, Drum Cleanout
59.	14-90	Nut, 5/16'' - 18 Hex.

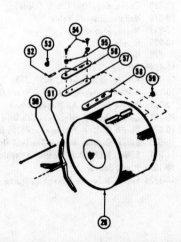

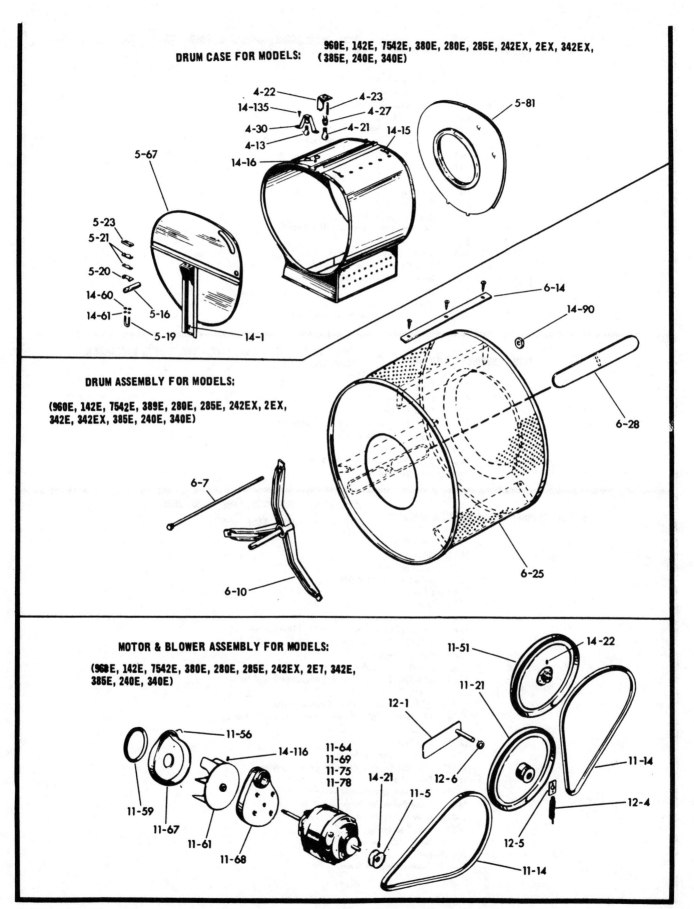

DRUM CASE FOR MODELS: 960E, 142E, 7542E, 380E, 280E, 285E, 242EX, 2EX, 342EX, (385E, 240E, 340E)

4-22
14-135
4-30
4-13
4-23
4-27
4-21
14-15
14-16
5-81
5-67
5-23
5-21
5-20
14-60
14-61
5-16
5-19
14-1
6-14
14-90
6-28

DRUM ASSEMBLY FOR MODELS:

(960E, 142E, 7542E, 389E, 280E, 285E, 242EX, 2EX, 342E, 342EX, 385E, 240E, 340E)

6-7
6-10
6-25

MOTOR & BLOWER ASSEMBLY FOR MODELS:

(960E, 142E, 7542E, 380E, 280E, 285E, 242EX, 2E7, 342E, 385E, 240E, 340E)

11-56
14-116
11-64
11-69
11-75
11-78
11-51
14-22
11-21
12-1
14-21
11-59
11-67
11-61
11-68
11-5
12-6
12-4
12-5
11-14
11-14

DRUM CASE FOR MODELS: (960E, 142E, 7542E, 380E, 280E, 285E, 242EX, 2E7, 342E, 342EX, 385E, 240E, 340E)

4-13	Germicidal Lamp (4 watt)
4-21	Flood Light (40 watt)
4-22	Flood Light Bracket
4-23	Tension Spring
4-27	Flood Light Socket Assembly
4-30	Germicidal Lamp Bracket & Socket Assembly
5-16	Main Bearing Assembly
5-19	Main Bearing U-Bolt Only
5-20	Main Bearing Clamp Angle
5-21	Main Bearing Shim
5-23	Main Bearing Filler Angle
5-67	Rear Drum Case Head
5-80	Thermostat Bulb Clip (Not Illustrated)
5-81	Front Drum Case Head
14-1	Screw, ½", #10 Round Head Type A
14-15	Drum Case Wrapper Sheet Angle Machine Screw
14-16	Drum Case Wrapper Sheet Angle Nut
14-60	Nut for U-Bolt
14-61	Washer for U-Bolt
14-135	Germicidal Bracket Attaching Screw

DRUM ASSEMBLY FOR MODELS: (960E, 142E, 7542E, 380E, 280E, 285E, 242EX, 2E7, 342EX, 385E, 240E, 340E)

6-7	Drum Tie Bolt
6-10	Drum Spider
6-14	Drum Tie Bar
6-25	Drum Assembly with Cleanout Door, Tie Bolts, But without Spider
6-28	Cleanout Cover
14-2	Drum Wrapper Sheet Metal Screws
14-90	Drum Tie Rod Nut

MOTOR BLOWER ASSEMBLY FOR MODELS: (960E, 142E, 7542E, 380E, 250E, 285E, 242EX, 2E7, 342E, 342EX, 385E, 240E, 349E)

11-5	Motor Pulley
11-14	Drive Belt
11-21	Intermediate Pulley
11-51	Drum Pulley (11-9, 960E)
11-56	Clips
11-58	Motor Saddle
11-59	Rubber Gasket
11-61	Exhaust Fan Assembly with #14-116 Setscrew
11-64	A-8762 1/6 HP Delco Motor with Throwout Switch
11-67	Fan Housing, Intake Side
11-68	Fan Housing, Exhaust Side
11-69	5KJ43DB7A G.E. 1/6 HP Aluminum Wound Motor with Throwout Switch
11-75	S60BKL-2536 1/6 HP Emerson Motor
11-78	1/6 HP Motor (G.E., Delco, Emerson) with Spade Connectors
12-1	Idler Bar
12-4	Idler Bar Tension Spring
12-5	Idler Bar Keyhole Plate for Tension Spring
12-6	Fiber Washer
14-19	Fan Housing Attaching Screw
14-21	Motor Pulley Set Screw
14-22	Drum Pulley Set Screw
14-57	Nut, Fan Housing
14-112	Motor Saddle Attaching Screw
14-113	Nut
14-116	Exhaust Fan Set Screw

DRUM, DRUM CASE, DRIVE ATTACHING PARTS & CONTROLS FOR MODELS:
(2E3, 3E9, 93E9, 4E9, 2E8, 3E8, 4E8, 1E1, 2E1,
3E1, 93E1, 4E1)

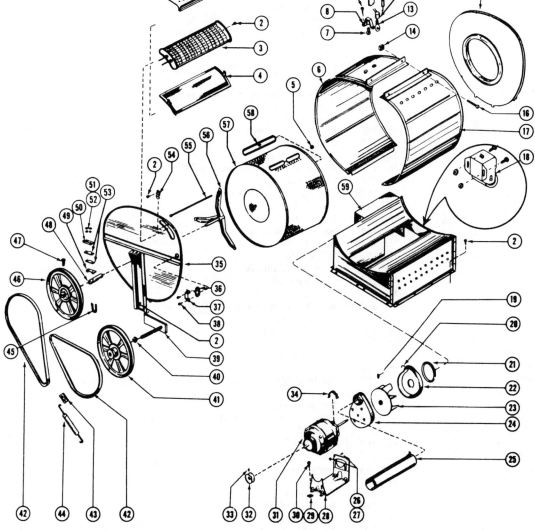

1.	8-21	Reflector, Heater
3.	8-1	Heater Assembly (With Screen)
4.	8-13	Screen, Heater
5.	14-90	Hex Nut, 5/16" - 18
6.	5-96	Wrapper, R.H. Drum Case—1E1, 2E1, 2E9
	5-71	Wrapper, R.H. Drum Case—3E1, 3E9
	5-105	Wrapper, R.H. Drum Case—93E1, 4E1, 93E9
7.	4-13	Bulb, "Sun E Day"—3E1, 3E8, 3E9
8.	4-30	Socket and Bracket, "Sun E Day"—3E1, 3E8, 3E9

**DRUM, DRUMCASE, DRIVE ATTACHING PARTS &
CONTROLS FOR MODELS:
(2E9, 3E9, 93E9, 4E9, 2E8, 3E8, 4E8, 1E1, 2E1, 3E1, 93E1, 4E1)**

9.	14-46	Screw, #6 Sheet Metal, 3/8"–3E1, 3E8, 3E9
10.	4-22	Bracket, Flood Lamp–3E1, 93E1, 3E8, 4E8, 93E1, 4E1, 3E9, 4E9, 93E9
11.	4-23	Spring, Flood Light Bracket–3E1, 93E1, 4E1, 3E8, 4E8, 3E9, 4E9, 93E9
12.	4-27	Socket, Flood Light–3E1, 93E1, 4E1, 3E8, 4E8, 3E9, 4E9, 93E9
13.	4-21	Bulb, Flood Light–3E1, 93E1, 4E1, 3E8, 4E8, 3E9, 4E9, 93E9
14.	14-16	Nut, Wrapper Sheet Angle, 5/16" - 24
15.	5-100	Head, Drum Case Front
17.	5-33	Wrapper, L.H. Drum Case
18.	9-89	Kit, Lint Screen Catch–3E1, 93E1, 4E1
20.	11-56	Clips, Fan Housing
21.	11-59	Gasket, Fan Housing
22.	11-82	Housing, Intake Fan (Alternate with 11-67)
23.	11-61	Fan, Exhaust
24.	11-83	Housing, Exhaust Fan (Alternate with 11-68)
25.	1-593	Pipe, Exhaust
28.	11-58	Saddle, Motor
29.	14-113	Nut, Saddle Attaching
31.	11-90	Motor, 1/6 H.P.
32.	11-5	Pulley, Motor
33.	14-21	Screw, Motor Pulley Set
34.	11-89	Clips, Motor attaching
35.	5-120	Head, Rear Drum Case
	5-67	Head, Rear Drum Case–2E9, 3E9, 93E9, 4E9, 2E8, 3E8, 4E8
36.	10-439	Protectomatic, Drum
37.	10-440	Shield, Protectomatic Fiber
38.	14-124	Screw, Sheet Metal, #8 - 5/8"
39.	12-7	Bar, Idler
	12-1	Bar, Idler–2E9, 3E9, 93E9, 4E9, 2E8, 3E8, 4E8
40.	12-10	Ring, Truarc
41.	11-93	Pulley, Intermediate
	11-21	Pulley, Intermediate–2E9, 3E9, 93E9, 4E9, 2E8, 3E8, 4E8
42.	11-14	Belts, Drive
43.	12-5	Plate, Keyhole
44.	12-4	Spring, Tension
45.	5-19	U-Bolt, Bearing
46.	11-51	Pulley, Drum
47.	14-22	Screw, Pulley Set
48.	5-16	Bearing, Main
49.	5-20	Angle, Bearing Clamp
50.	5-23	Angle, Bearing Filler
51.	14-60	Hex Nut, U-Bolt, 5/16" - 18
52.	14-61	Washers, U-Bolt
53.	5-21	Shim, Bearing
54.	8-11	Clip, Heater Fastening
55.	6-7	Bolts, Drum Tie
56.	6-10	Spider, Drum
57.	6-25	Drum Assembly–2E1, 3E1, 93E1, 4E1, 2E9, 3E9, 93E9, 4E9, 2E8, 3E8, 4E8
	6-36	Drum Assembly–1E1
58.	6-28	Cover, Drum Cleanout–2E1, 3E1, 93E1, 4E1, 2E9, 3E9, 93E9, 4E9, 2E8, 3E8, 4E8
	6-37	Cover, Drum Cleanout–1E1
59.	7-62	Housing, Air Chamber–1E1, 2E1
	7-65	Housing, Air Chamber–3E1, 93E1, 4E1

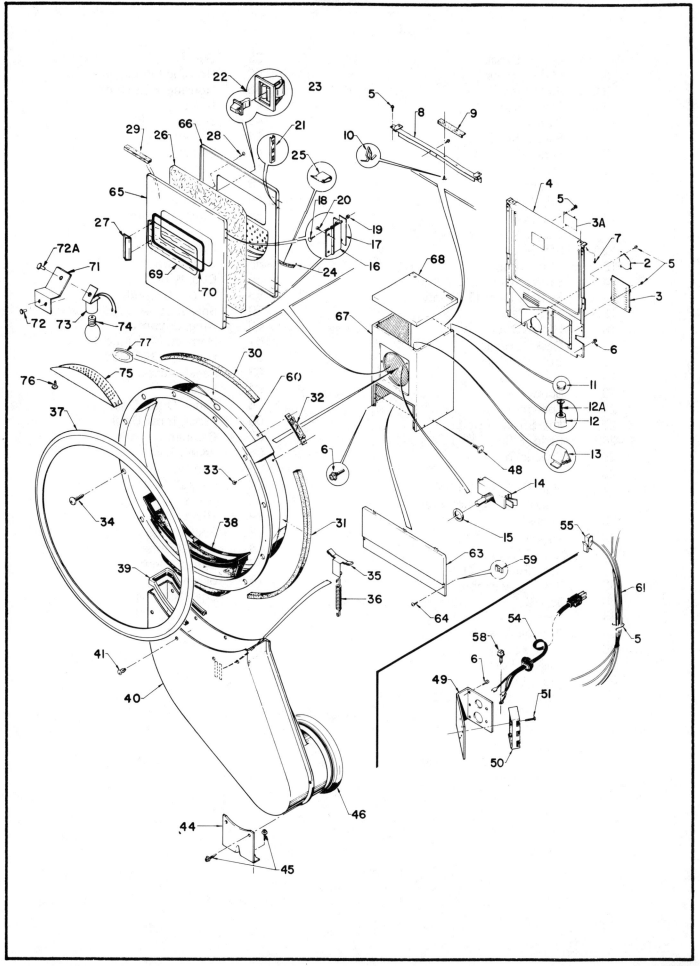

2	Cover, Exhaust
3	Plate, Access
3A	Plate, Access
4	Panel, Rear
5	Screw
6	Screw, Sems
7	Screw
8	Strut
9	Shield
10	Clip, Dart
11	Bumper
12	Guide
12A	Screw
13	Clip, Spring
14	Switch, Door (453-253)
14	Switch, Door (353 only)
15	Nut , Hex
16	Hinge, Door
17	Gasket, Hinge
18	Bolt, Carriage
19	Nut
20	Screw
21	Speednut
22	Kit, Strike & Catch
24	Strip, Filler
25	Speedclip
26	Foil
27	Handle, Door (White)
27	Handle, Door (Chrome)
28	Screw
29	Spacer
30	Seal, Felt (Short)
31	Seal, Felt (Long)
32	Bearing Assembly
33	Screw
34	Screw
35	Bearing Assembly
36	Spring
37	Gasket, Port
38	Trap, Lint

39	Gasket
40	Housing (353 & 253)
40	Housing (453 Only)
41	Rivet
43	Speedclip (453 Only)
44	Bracket
45	Screw
46	Gasket
48	Screw
49	Bracket
50	Block, Terminal
51	Screw
54	Cord, Service
55	Clip, Harness
56	Clip, Wire
58	Screw, Ground
59	Speednut
60	Ring, Support
61	Harness, (453 Models)
61	Harness, (353 Models)
63	Panel, Access
64	Screw
65	Panel, Outer Door
66	Panel, Inner Door
67	Cabinet
68	Panel, Top
69	Window
70	Gasket, Seal
72	Rivet
72A	Rivet
73	Socket
74	Bulb, Ozone
75	Shield, Light
76	Screw
77	Snap Bushing

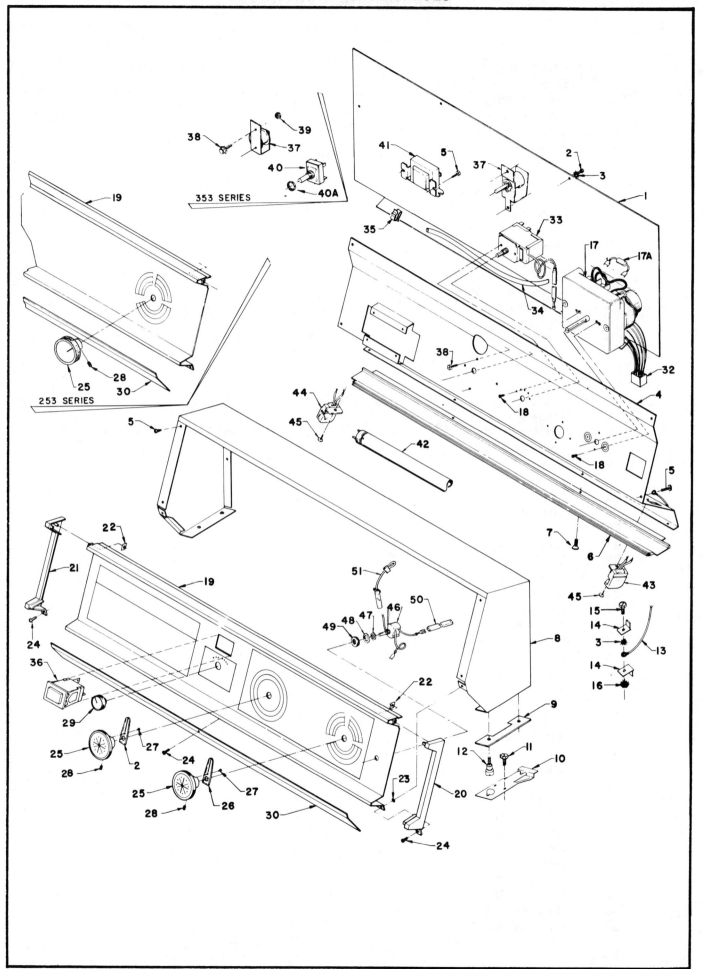

353 SERIES

253 SERIES

HOUSING AND CONTROLS

1	Backplate
2	Screw
3	Lockwasher
4	Bracket, Mounting
5	Screw
6	Extrusion
7	Screw
8	Housing (White)
8	Housing (Coppertone)
8	Housing (Avocado)
8	Housing (Harvest Gold)
9	Spacer
10	Clip, Spring
11	Screw
12	Screw, Shoulder
13	Wire, Ground
14	Terminal, Ground
15	Screw, Sems
16	Nut, Keps
17	Timer
17A	Resistor
18	Screw
19	Trimplate
20	End Cap, R.H.
21	End Cap, L.H.
22	Speednut
23	Lockwasher
24	Screw
25	Knob Assembly
26	Pointer
27	Screw, Drive
28	Set Screw
29	Knob
30	Lens
32	Harness
33	Thermostat
34	Sleeve, Capillary
35	Clip, Capillary
36	Switch, Start
37	Buzzer
38	Screw, Sems
39	Nut, Keps
40	Switch, Rotary
40A	Nut, Hex
41	Ballast
42	Lamp, Fluorescent
43	Lampholder, R.H.
44	Lampholder, L.H.
45	Screw
46	Switch, Light
47	Nut, Hex
48	Lockwasher
49	Nut, Knurled
50	Connector, Wire
51	Wire, Jumper (12")

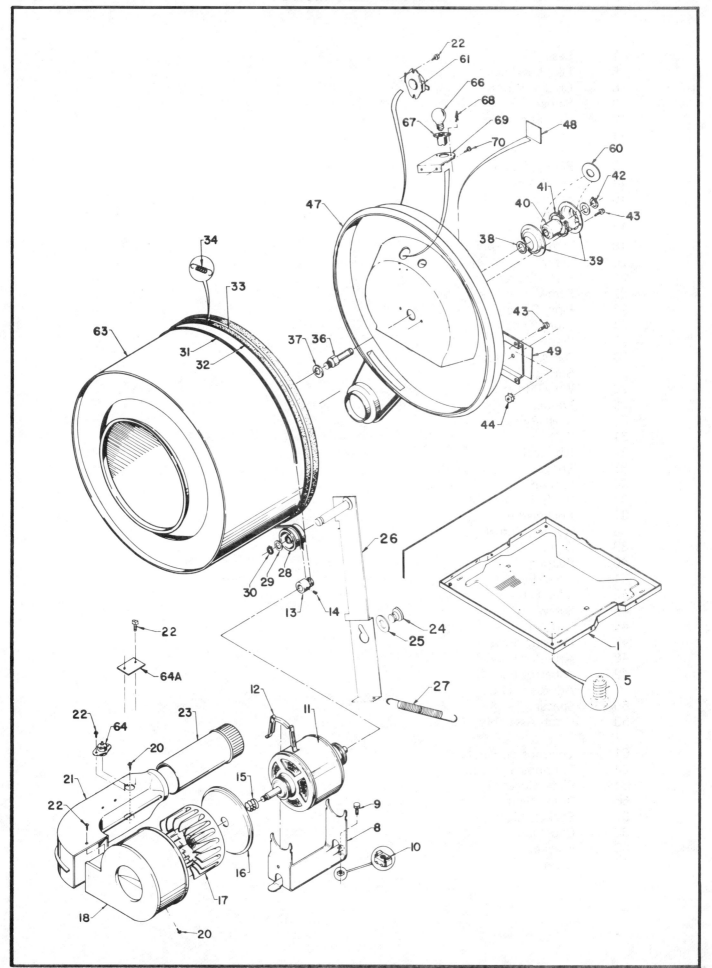

DRUM AND BLOWER

1	Base
5	Leg, Leveling
8	Cradle, Motor
9	Screw
10	Nut, Speedgrip
11	Motor
12	Clip
13	Pulley, Motor
14	Set Screw
15	Spring
16	Cover
17	Impeller
18	Housing, Blower
20	Screw
21	Duct, Outlet
22	Screw
23	Pipe, Outlet
24	Stud, Pivot
25	Washer
26	Bar Assem., Idler
27	Spring
28	Wheel, Idler
29	Washer
30	Snap Ring
31	Belt
32	Kit, Seal
33	Strap
34	Spring
36	Shaft, Drum
37	Lockwasher
38	Washer, Thrust
39	Collar
40	Bearing
41	Wick, Felt
42	Snap Ring
43	Screw
44	Nut
47	Housing, Heat
48	Asbestos (3x4½)
49	Asbestos (6x4½)
*	Adhesive (1 oz.)
60	Spacer
63	Drum Assembly
61	Thermostat, Safety
64	Thermostat, Control
64A	Plate, Cover
66	Bulb, Light
67	Socket, Light
68	Clip, Dart
69	Bracket
70	Rivet

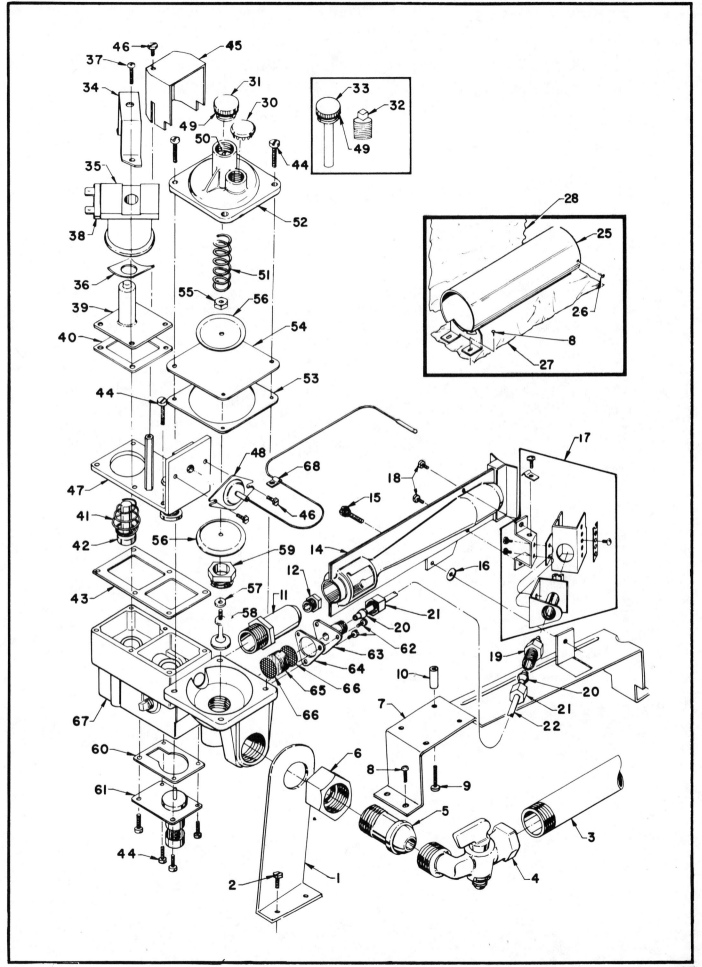

BURNER

1	Bracket, Support	35	Solenoid
2	Screw	36	Spring
3	Pipe, Inlet	37	Screw
4	Valve, Shutoff	38	Insulator, Terminal
5	Tail Piece	39	Plate Assembly
6	Nut, Union	40	Gasket
7	Bracket, Mounting	41	Spring , Plunger
8	Screw	42	Plunger, Assembly
9	Screw	43	Gasket
10	Spacer	44	Screw
11	Holder, Orifice	45	Cover
12	Orifice, Main (Nat., Mix.)	46	Screw
12	Orifice, Main (Mfd.)	47	Kit, Plate & Latch
12	Orifice, Main (L.P.)	48	Mercury Element
14	Burner, Main (Nat., Mix.)	49	Gasket
14	Burner, Main (L.P.)	50	Screw, Regulator
15	Screw	51	Spring, Reuglator
16	Washer	52	Cover, Regulator
17	Burner, Pilot	53	Gasket
18	Screw	54	Diaphragm, Regulator
19	Orifice, Pilot (Nat., Mix.)	55	Hex Nut
19	Orifice, Pilot (Mfd.)	56	Plate, Diaphragm Support
19	Orifice, Pilot (L.P.)	57	Washer
20	Sleeve, Compression	58	Plunger, Regulator
21	Nut, Compression	59	Nut, Brass
22	Tube, Pilot	60	Basket
25	Chamber, Combustion	61	Plunger Asembly
26	Screw	62	Screw
27	Asbestos (9 x 15)	63	Fitting Assem., Pilot
28	Asbestos (16 x 18)	64	Gasket
30	Plug, Vent (Nat., Mix., Mfd.)	65	Filter, Pilot
31	Cap, Regulator (Nat., Mix., Mfd.)	66	Screen, Filter
32	Pipe Plug, Vent (L.P.)	67	Body, Valve
33	Plug, Block Open (L.P.)	68	Clamp, Capillary
34	Bracket, Solenoid		

NOTE: The following Gas Conversion Kits are available:

1. For Nat., Mix., Mfd. to L.P. use Kit No. F76885-1.

2. For L.P. to Nat., Mis., Mfd. use Kit No. F76886.

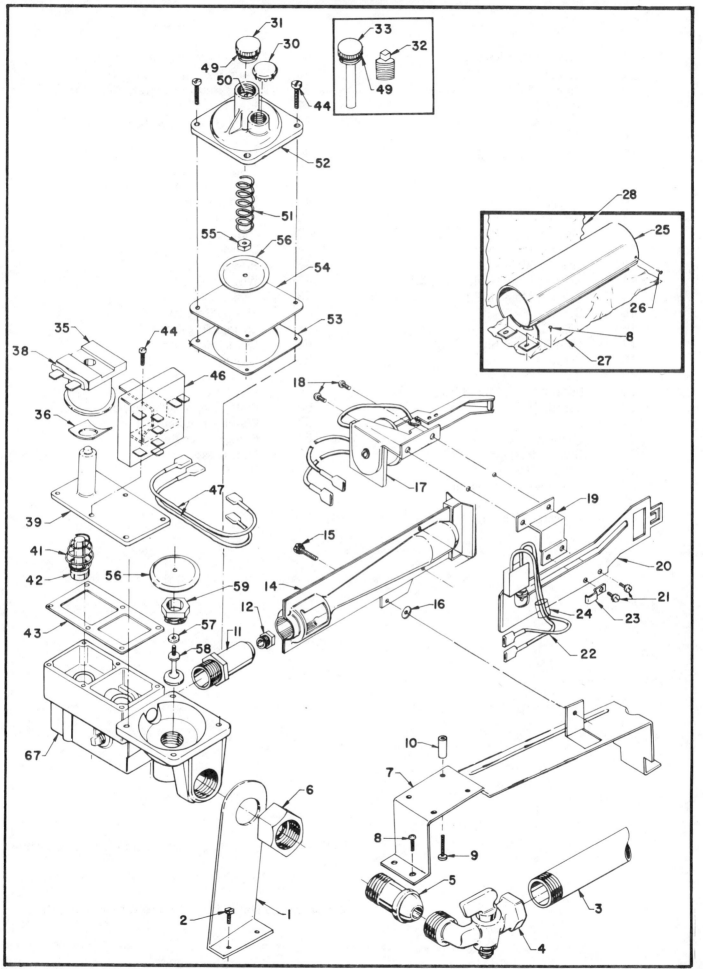

BURNER

1	Bracket, Support		28	Asbestos (16 x 18)
2	Screw		30	Plug, Vent (Nat., Mix., Mfd.)
3	Pipe, Inlet		31	Cap , Regulator (Nat., Mix., Mfd.)
4	Valve, Shutoff		32	Pipe Plug, Vent (L.P.)
5	Tail Piece		33	Plug, Block Open (L.P.)
6	Nut, Union		35	Solenoid
7	Bracket, Mounting		36	Spring
8	Screw		38	Insulator, Terminal
9	Screw		39	Plate Assembly
10	Spacer		41	Spring, Plunger
11	Holder, Orifice		42	Plunger Assembly
12	Orifice, Main (Nat., Mix)		43	Gasket
12	Orifice, Main (Mfd.)		44	Screw
12	Orifice, Main (L.P.)		46	Power Pack
14	Burner, Main (Nat., Mix., Mfd.)		47	Wire, Jumper
14	Burner, Main (L.P.)		49	Gasket
15	Screw		50	Screw, Regulator
16	Washer		51	Spring, Regulator
17	Ignitor Assembly		52	Cover, Regulator
18	Screw		53	Gasket
19	Bracket, Sensor		54	Diaphragm
20	Sensor (Nat., Mix., Mfd.)		55	Hex Nut
20	Sensor (L.P.)		56	Plate, Diaphragm Support
21	Screw		57	Washer, Rubber
22	Harness		58	Plunger, Regulator
23	Clamp, Harness		59	Nut, Brass
24	Sleeve, Plastic		67	Body
25	Chamber, Combustion			
26	Screw			
27	Asbestos (9 x 15)			

NOTE: The following Gas Conversion Kits are available:

1. For Nat., Mix., Mfd. to L.P. use Kit No. F76885-1.

2. For L.P. to Nat., Mis., Mfd. use Kit No. F76886.

WIRING DIAGRAM SYMBOLS

The first column below indicates the wiring diagram item; the second column contains symbols currently in use; the third column contains the wiring diagram symbols to be used in future production.

Item	Current Std.	Revised Std.
Motor Single Speed		
Motor Multi Speed	1725 1140	1725 1140

NOTE: Internal motor wiring may be shown

ITEM	CURRENT LDRY. STD.	REVISED LDRY. STD.
Internal Conductor		
Harness Wire		
Permanent Connection		
Cross Over		
Ground		
Timer Switch		
Automatic Switch		
Manual Switch		
Double Throw		
3 Prong Plug		
Heater (Wattage Shown)	2800	2800W
Fuse		
Circuit Breaker		
Terminal		
Timer Motor		
Thermister	NONE	
Transistor	NONE	
Diode (Rectifier)	NONE	

Item	Current	Revised
Light (Incandescent)		
Germ Lamp		
Pressure Switch		
Fluorescent		
Starter (Automatic)	S	S
Coil		
Capacitor		
Resistor (Show Value)	500	500Ω
Plug Connector		
Centrifugal Switch		
Thermostat Show N.O. or N.C.		
Double Throw Stat		
Ballast		B
Adj. Stat		
Thermocouple		
Warp Switch		
Neon Light	NONE	
Transformer	NONE	
Rectifier (Controlled)	NONE	
Relay	Coil & Switches separate in circuit.	

HERE'S A HANDY CHART FOR FIGURING ELECTRICAL VALUES

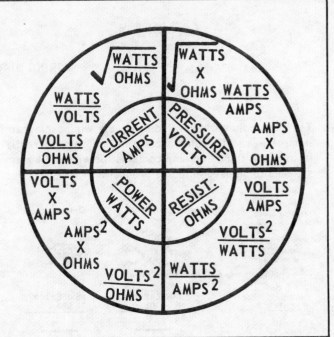

This chart can be used as a quick reminder of electrical formulas. It shows 3 ways to figure each value, AMPS, VOLTS, OHMS or WATTS and can be easily copied and carried in the shirt pocket or cemented to the inside of the tool box.

Memorandum

_____

_____

_____

_____

_____

_____

_____

_____

_____

NOTES

REPAIR-MASTER®
For AUTOMATIC WASHERS, DRYERS & DISHWASHERS

These Repair-Masters offer a quick and handy reference for the diagnosis and correction of service problems encountered on home appliances. They contain many representative illustrations, diagrams and photographs to clearly show the various components and their service procedure.

Diagnosis and repair charts provide step-by-step detailed procedures and instruction to solve the most intricate problems encountered in the repair of washers, dryers and dishwashers.

These problems range from timer calibrations to complete transmission repair, all of which are defined and explained in easy to read terms.

The Repair-Masters are continually updated to note the latest changes or modifications in design or original parts. These changes and modifications are explained in their service context to keep the serviceman abreast of the latest developments in the industry.

AUTOMATIC WASHERS

9001 Whirlpool-Kenmore	9013 Philco-Bendix Top Loader	5556 Whirlpool
9002 Norge and Hamilton	9014 Franklin	5557 Frigidaire
9003 General Electric	9015 Norge - Plus Capacity	
9003-S General Electric (Spanish)	9016 Frigidaire Roller-Matic	**CLOTHES DRYERS**
9004 Hotpoint	9017 Westinghouse Top Loading	
9005 Westinghouse Front Loading		8051 Whirlpool-Kenmore
9006 Kelvinator	**DISHWASHERS**	8052 General Electric
9008 Easy		8053 Hamilton
9009 Frigidaire	5551 Hotpoint	8054 Norge/Fedders
9010 Maytag	5552 General Electric	8055 Maytag
9011 Philco-Bendix Front Loading	5553 Kitchenaid	8056 Westinghouse
9012 Speed Queen	5554 Westinghouse	8057 Speed Queen
	5555 D & M	8058 Franklin

* The D & M Dishwasher Repair-Masters covers the following brands:
Admiral — Caloric — Chambers — Frigidaire (Portable) — Gaffers and Sattler — Gibson — Kelvinator — Kenmore — Magic Chef — Magic Maid — Norge — Philco — Pioneer — Preway — Roper — Wedgewood — Westinghouse (Portable)

The Repair-Master® series will take the guesswork out of service repairs.

REPAIR MASTER® for GAS BURNER CONTROLS

For Domestic GAS DRYERS

* DIAGNOSIS CHARTS

* CHECKING PROCEDURE

* SERVICE PROCEDURE

* COMPONENT DATA

6001

92 Pages

A most comprehensive look at gas burners, and controls, their problems, diagnosis and repair.

Showing adjustments, service procedures, illustrations, wiring diagrams, and photos, related to domestic gas dryer service.

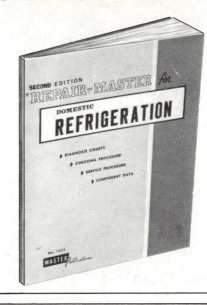

REPAIR MASTER® For WINDOW AIR CONDITIONERS

A Complete Service and Repair Guide!

The concise, easy to read format will enable you to clearly understand the functions of the various components and aid you in properly servicing them.

7541
80 Pages

This Repair-Master® offers a quick and handy reference for the diagnosis and correction of service problems. Many illustrations, diagrams and photographs are included to clearly show the various components and their service procedures.

PRINCIPLES OF REFRIGERATION

* DIAGNOSIS CHARTS
* TEST CORD TESTING
* PROCEDURES
* STEP-BY-STEP REPAIR
* SHOP PROCEDURES

REPAIR MASTER® For DOMESTIC AUTOMATIC ICEMAKERS

* DIAGNOSIS CHARTS

* TESTING PROCEDURES

* ADJUSTMENT PROCEDURES

* STEP-BY-STEP REPAIRS

7531
176 pages

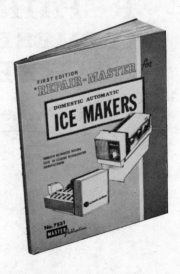

Includes ten different designs of Automatic Ice Makers presently found in domestic refrigerators.

A special section is included for Whirlpool design self-contained Ice Maker. Illustrations, charts and new parts design information makes this Repair-Master® an important tool.

TECH-MASTER® For REFRIGERATORS & FREEZERS

Handy 8½" x 5½" book is designed to be easily carried in the tool box or service truck. The first time you use these books they will more than repay the nominal purchase price.

Operating Information for Over 12,900 Models

* TEMPERATURE CONTROL PART NUMBERS
* CUT-IN AND CUT OUT SETTINGS
* UNIT HORSEPOWER
* REFRIGERANT AND OIL CHARGE IN OUNCES

For All Major Brands

Offering a quick reference by brand, year, and model number.

No. 7550	No. 7550-2	No. 7550-3
1950-1965	1966-1970	1971-1975
429 pages	377 Pages	332 pages

* SUCTION AND HEAD PRESSURES
* START AND RUN WATTAGES
* COMPRESSOR TERMINAL HOOK-UPS
* DEFROST HEATER WATTAGES